The National Computing Centre develops techniques, provides services, offers aids and supplies information to encourage the more effective use of Information Technology. The Centre co-operates with members and other organisations, including government bodies, to develop the use of computers and communications facilities. It provides advice, training and consultancy; evaluates software methods and tools; promotes standards and codes of practice; and publishes books.

Any interested company, organisation or individual can benefit from the work of the Centre – by exploring its products and services; or in particular by subscribing as a member. Throughout the country, members can participate in working parties, study groups and discussions; and can influence NCC policy.

For more information, contact the Centre at Oxford Road, Manchester M1 7ED (061-228 6333), or at one of the regional offices: London (01-353 4875), Bristol (0272-277 077), Birmingham (021-236 6283), Glasgow (041-204 1101) or Belfast (0232-665 997).

Do You Want to Write?

Could you write a book on an aspect of Information Technology? Have you already prepared a typescript? Why not send us your ideas, your 'embryo' text or your completed work? We are a prestigious publishing house with an international reputation. We have the funds and the expertise to support your writing ambitions in the most effective way.

Contact: Geoff Simons, Publications Division, The National Computing Centre Ltd, Oxford Road, Manchester M1 7ED.

What is EDI?

Martin Preston

PUBLISHED BY NCC PUBLICATIONS

British Library Cataloguing in Publication Data

Preston, Martin
  What is EDI?
  1. Computer-telecommunication services
  I. Title
  384.3

  ISBN 0-85012-732-7

© THE NATIONAL COMPUTING CENTRE LIMITED, 1988

All rights reserved. No part of this publication may be reproduced, stored in a retrieval system, or transmitted, in any form or by any means, without the prior permission of The National Computing Centre.

First published in 1988 by:

NCC Publications, The National Computing Centre Limited, Oxford Road, Manchester M1 7ED, England.

Typeset in 10pt Melior by Bookworm
and Printed by Hobbs the Printers of Southampton.

ISBN 0-85012-732-7

Acknowledgements

I am grateful to the following people who kindly offered to review this publication and provided valuable comments:

Tony Metcalf of Philips Electronics and the EDI Association.

Keith Blacker of Lucas Engineering and Systems Limited and the EDI Association.

Nigel Fenton of the Article Number Association.

Paul Chilton of the National Computing Centre Limited.

Contents

Page

Acknowledgements

1 Introduction — 1

Data Interchange Problems — 1
Movement Towards EDI — 2
Future EDI Developments — 2
Aims of the Book — 2

2 The Background to EDI — 3

3 What is EDI? – Definition and Types of Information — 7

The Problems of Definition — 7
EDI Versus Electronic Mail — 9

4 What Does EDI Provide? – The Business Case and Benefits — 11

Introduction — 11
The Benefits of EDI — 12
The Impact on Organisations — 16
Assessment of Risk — 18
Conclusion — 21

5 What Does EDI Look Like? – Sample Scenario — 23

In the Beginning ... — 23
Then Came EDI ... — 24
And Now ... — 25
But, Back to Reality ... — 25

6 Communications Issues — 27

Transmission Media — 27
VADS — 29
The Enabling Software — 35
Conclusion — 37

7 Standards — 39

What Are EDI Standards? — 39
Why Are EDI Standards Important? — 40
What Form do EDI Standards Take? — 41
How Have EDI Standards Developed? — 42
What Next for Standards? — 45

8 Planning The Implementation — 47

Preliminary Steps — 47
Planning Towards EDI — 48
What Next After The Pilot? — 50

9 UK EDI Applications — 53

Automobile — 53
Consumer Goods — 54
Ports — 54
Airports — 55
HM Customs and Excise — 55
Transport and Shipping — 56
Chemicals — 57
Electronics — 57
Health — 57
Banking — 58
Insurance — 59
Construction — 59
Tourism — 60

10 International Issues — 61

1992 — 61
European Initiatives — 61

11 Security and Legality 67

How Can I Be Sure of the Sender's Identity? 67
Can I Be Assured of the Integrity of the Message? 68
Is the Transmission Mechanism Secure and Confidential? 68
How Can The Sender Be Sure that the Message Arrived? 69
Does the EDI Message Constitute a Binding Contract? 69
What is the Legal Standing of an EDI Transaction? 70
Conclusion 70

12 Conclusion and Future Developments 73

Appendix

1 Glossary 77

2 EDI Groupings and Contacts 89

Index 95

1 Introduction

EDI, **E**lectronic **D**ata **I**nterchange, is the name given to the exchange of structured trade data between the computer systems of trading partners. It is predicted that this method of 'paperless' trading will revolutionise the way companies do business.

DATA INTERCHANGE PROBLEMS

In simple terms EDI refers to the transmission and receipt of such commonplace trading documentation as invoices, purchase orders and credit notes. In the non-EDI world, the exchange of this information has usually involved the keying and re-keying of data from paper documents into computer-based systems, with the necessary checking, printing and packaging that this entails, not to mention the vagaries of the postal system. The delays, errors and costs inherent in this process will be familiar to all the parties involved. The impact on business is equally familiar, as the flow of goods is restrained by slow-moving paperwork and manufacturing is forced to hold large inventories. The impact on the customer is invariably the higher cost of goods or services. Today's computer and telecommunications technology provides a viable solution to these problems. Chapters 2 to 5 of this book cover the background, definition and benefits of EDI.

Many companies are already using EDI to solve these problems and are aware of the benefits of this approach. In the UK this covers a wide range of business sectors, including retail, manufacturing, transport, construction, shipping, finance and the public sector. National EDI developments are well advanced in the US and Europe. EDI is, however, truly international, reflecting the nature of today's

2 WHAT IS EDI?

trading, and encompassing forwarders, carriers, HM Customs and Excise, and governments. The spread of EDI across national barriers is being augmented by the development of international standards for documentation, as discussed in Chapter 7. Chapters 9 and 10 provide an overview of the current practice of EDI.

MOVEMENT TOWARDS EDI

There is no doubt that EDI is being driven forward by the demands of the business user community. Fuelled by this demand, the computer industry has not been slow to play its part, by providing networks and associated 'mailing' services, and by developing the tools to enable users to link their computer systems directly to the network. The supplier marketplace now includes such established names as IBM, DEC, ICL, General Electric, Istel and Systems Designers. The network service industry has also spawned joint venture companies such as INS, where ICL and GEISCO have combined forces to offer a multi-national network covering over 70 countries. Chapter 6 covers the vendor services.

FUTURE EDI DEVELOPMENTS

EDI is growing extremely rapidly and many forecasts and predictions exist for the likely level, cost, and indeed, the nature of trading in the future. These figures are almost always prefaced by the word 'conservative'. No doubt there are problems yet to be overcome before EDI can be said to have totally revolutionised the trading cycle, not least being the need for legal changes as discussed in Chapter 11. These changes are being effected and the question regarding the future scenario, is not 'if?', but 'when?'. Chapter 12 looks at the future developments of EDI.

AIMS OF THE BOOK

This book provides an introduction to EDI. The intention is not to provide a thorough treatment of any particular aspect of the subject, but to explain the background, concepts, technologies and issues, and to illustrate this with reference to practical user experience. Like so many new developments EDI has grown its own brand of jargon; this book should help to break down the jargon and explain the underlying importance of this exciting subject.

2 The Background to EDI

Today's commercial environment is highly volatile and competitive. Businesses can no longer rely upon a fixed and stable market for their products, and must compete in terms of service, quality and cost against an ever increasing list of rivals. This not only implies stringent cost-cutting but, perhaps more significantly, changes to the methods of manufacturing, trading relationships, and the very culture of organisations. One example is the adoption by many manufacturing companies of 'just-in-time' techniques, characterised by a reduced manufacturing cycle time, machine and worker flexibility, and minimal stock holding. In this environment EDI is a powerful weapon which organisations ignore at their peril.

EDI is not new. Its origins may be traced back to the United States when in the 1960s various industry sectors (airlines, car manufacturing and health) established EDI trials, though in comparison with today's technology, involving poorly developed standards and inflexible communications links. The first practical application of EDI is generally regarded as being the LACES (London Airport Cargo EDP Scheme) system for cargo clearance at London's Heathrow airport. Introduced in the early 1970s this system was successful in convincing the sceptics that security and confidentiality were not insurmountable issues. Indeed it was demonstrated that EDI offered much improved methods of handling business data. However, important as LACES was, several developments have been crucial to the widespread adoption of EDI.

- *The computer revolution.* As computer systems began to proliferate in the 1970s prompted by cheap, user friendly and versatile hardware and software, and competition amongst suppliers, so businesses became accustomed to using these systems for local processing. Multipurpose office workstations have increasingly become commonplace.

- *The telecommunications revolution.* The enhancement of techniques for transmitting digital data, including the advent of packet switching networks, electronic exchanges, local and wide area networks, and optical fibre has enhanced both the quality and quantity of data transmission. It is now commonplace to talk of a single computer/telecommunications revolution, as companies have moved towards distributed computer processing and intra-company data transfer.

- *The deregulation of telecommunications.* This revolution has been particularly significant to UK EDI developments because of the liberalisation of the UK telecommunications industry. This deregulation has led to the emergence of a number of third-party networks, in some cases offering alternative communications facilities to what had previously been a domestic monopoly, eg Mercury in the UK.

- *An awareness of benefits.* A 1984 survey, conducted jointly by Istel Ltd and the Institute of Physical Distribution Management (IPDM) on the costs of generating and processing documents, found that commercial documents cost £10 each to process, and over two thirds of this cost was associated with people and paper. Similar studies have highlighted these administrative cost savings and have contributed to the first phase in raising the level of awareness. The Istel/IPDM survey results showed that over two thirds of the UK companies surveyed thought that their organisation would benefit from using EDI (see Figure 2.1 reproduced with permission from Istel Ltd). The business community is now being subjected to the second phase of benefit awareness, the benefits being more intangible, but perhaps more significant. These may include the facilitation of new manufacturing methods and a greater understanding of the nature of trading partner operations.

THE BACKGROUND TO EDI

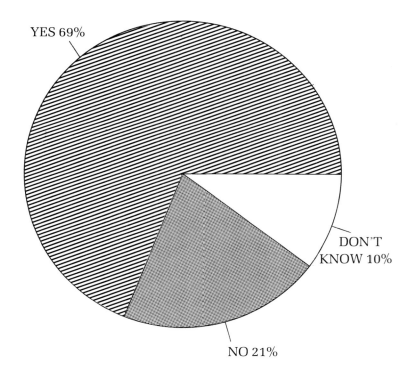

Figure 2.1 Would Organisations Benefit From Using EDI?

3 What is EDI? – Definition and Types of Information

THE PROBLEMS OF DEFINITION

Anyone browsing through the various EDI publications and press articles looking for a precise and simple statement of what constitutes Electronic Data Interchange, will almost certainly encounter a range of definitions. Each definition will have been put forward in an attempt to encapsulate the essential features of EDI, whilst trying to exclude (or in some cases even include) the other related technologies. The International Data Exchange Association (IDEA) provides the following definition:

> The transfer of structured data, by agreed message standards, from one computer system to another, by electronic means.

In essence this statement simply refers to the much-used phrase 'paperless trading', the freeing of business from the mountain of administrative paperwork. However, it is worthwhile examining some of the terminology used in this definition.

"Structured data" refers to a precise, recognised and accepted method of assembling data. Such data items as product number, customer name, and unit price may be structured into a purchase order or invoice for example. The very name 'invoice' conjures up a picture of the data that we expect to receive. We may contrast EDI with electronic mail where the equivalent data may be transmitted in the form of an ad hoc enquiry, containing no recognised form. Several definitions of EDI refer to 'structured trade data', which is perhaps more useful in that it focuses the mind on commonplace business documentation: invoice, credit note, purchase order, packing list, acknowledgement of order, etc. Indeed the phrase Trade Data Interchange has been frequently used to mean Electronic Data Interchange. EDI, however, also encompasses the exchange of engineering product design data – Product Data Interchange, interactive enquiry/confirmation messages, and electronic funds transfer.

A cursory examination of the invoices provided by two different companies will undoubtedly highlight differences. The customer name and address may vary in its position on the invoice, the date may be provided in different formats, or descriptive text may or may not be provided. The phrase "by agreed message standards" implies that such discrepancies between invoices (an invoice is one such message) will be minimised by providing a fixed and agreed method of specifying and presenting the data. Much effort has been expended by respected national and international bodies in producing standards for specifying the data, via a data directory, and standards for presenting the data, via syntax rules and message guidelines; this subject is explored further in Chapter 7.

The definition uses the phrase "from one computer system to another", implying that the two systems belong to distinct organisations. This implication is not stated directly in the definition and readers may wish to include both intra-company and inter-company communications, provided that the transactions are between trading partners, usually between supplier and customer. This definition also specifically excludes the passing of information from a data entry terminal attached to a host computer: two machines exist but only one computer system. In this respect, the practice of Direct Trader Input (DTI) is becoming commonplace in the shipping industry, where carriers and agents may send customs entries direct from on-line terminals to the customs computer. DTI is certainly in keeping with the spirit and aims of EDI, and certainly involves two separate organisations exchanging data, but argument may be raised over whether it should be covered by the EDI umbrella.

In this same category may also be included dumb terminal-to-computer proprietary systems, where communications links are provided from a company computer to remote users for purpose of order placement and confirmation; again, not a computer-to-computer system and so not included within the term EDI.

WHAT IS EDI? – DEFINITION AND TYPES OF INFORMATION 9

The phrase in the definition "from one computer system to another, by electronic means", implies a direct computer-to-computer link for electronic data transfer. EDI is certainly striving to achieve this and without doubt this linkage is essential to fully realising the benefits offered by EDI. However, much 'paperless trading' is currently practised using magnetic tape as the transfer mechanism, and this form of EDI is certainly included within the definition.

The definition makes no reference to the timing factors involved in EDI, although it should be apparent that the trading cycle will be dramatically altered. Most current EDI applications use some form of batch transfer mechanism, whereby the data is stored, prior to forwarding on to the recipient. However, some business transactions require a more immediate and conversational method of exchanging information, whereby the two parties actively cooperate at the time of the exchange of data. An example of this is in the travel and leisure industry, where a holiday booking may necessitate a 'conversation' between tour operator, travel agent and ferry company in order to ensure confirmation of the holiday. This interactive EDI is a facet of 'paperless trading' which is, as yet, not as fully developed as the batch, or store-and-forward, method.

EDI VERSUS ELECTRONIC MAIL

It can probably now be appreciated that EDI covers a wide variety of business applications. Many readers may already be familiar with the concept of electronic mail as a method of exchanging information and, as a further aid to clarity, the essential differences between these two technologies are now itemised.

- *Interchange agreement.* EDI trading commonly involves an agreement between trading partners concerning the types of information to be transmitted and providing legal status to the electronic documents. Even where no formal agreement exists, the EDI partners have a much higher level of expectation of what will appear in the electronic mailbox and when, than do the electronic mail users.

- *Structured data.* As previously stated, EDI is concerned with specific presentations of trade data, to specific standards. Although data may be structured for electronic mail, for example in telex systems, this is not a cornerstone of the technology as it is in EDI.

— *Personal/company mailbox.* EDI systems are likely to have company or functional electronic mailboxes (eg invoicing, order processing) as opposed to the personal mailboxes prevalent in electronic mail systems. This distinction is associated with the necessity for user intervention when processing electronic mail, whereas the basis of EDI is that the mail may be processed automatically by computer systems, ie the rules for processing invoices, purchase orders, etc, are fixed and programmable.

In summary, EDI may be seen as a cooperative system, requiring the collaboration of at least two parties, usually with different business objectives, to form a joint computer-based trading system. These cooperative systems may thus involve corporations and their suppliers and customers, corporations and their banks, companies in joint ventures with each other, or even competitors. Why should companies take this radical step forward? In answer to this question, the business case and benefits are discussed in Chapter 4.

4 What Does EDI Provide? – The Business Case and Benefits

INTRODUCTION

For many, if not all businesses, EDI is a radical step forward. It involves not only an investment in computing hardware and software, network charges and development costs, but, perhaps more significantly, a re-think of the relationship between a customer and supplier, an adjustment in working practices, and a dramatic alteration to the trading cycle. The casual onlooker may anticipate that in such circumstances a 'hard sell' approach would be required from vendors and consultants, pushing hard against an in-built force of resistance from the user community. However, this is far from the true picture. The impetus for EDI has come as much from the user community as from the vendor industry itself. Indeed, in many cases the vendor industry may be blamed for resisting the development of EDI, being slow to develop suitable enabling hardware and software and to provide the requisite links between communications networks.

This push from the user community is underpinned by a firm appreciation of the benefits offered by EDI, benefits which provide not only departmental operational savings (eg administrative cost reductions), but also benefits of strategic significance to the corporate body, and benefits in enhanced business opportunities.

This chapter will examine the strategic, operational and opportunity benefits offered by EDI. This assessment of benefits is an essential process for any organisation to undertake if it is considering implementing EDI. Indeed, the organisation will need to produce methods of quantifying these benefits, and hence to justify EDI against the costs, requirements and alternative solutions. However,

12 WHAT IS EDI?

this quantification is outside the scope of this publication, as is the method of evaluating alternative EDI options.

Before embarking on a discussion of benefits, and in order to strike a point of balance, it should be pointed out that for many companies the push to implement EDI in order to realise countless benefits will not represent their own track record. These companies may well have been forced into EDI by the insistence of a dominant customer as a condition of contract for continuing business, and they may see the benefits as being weighted firmly towards the customer. However, even in such circumstances the supplier should be quick to see the competitive edge that EDI can provide in offering such a favourable supply mechanism to future customers.

THE BENEFITS OF EDI

The benefits of EDI may be conveniently separated into strategic, operational, and opportunity, where the following meanings apply:

- *Strategic*. Of crucial, long-term significance to the functioning of the organisation. These benefits will affect the very business the company is undertaking, ie its central operating function.

- *Operational*. Of crucial significance to the daily operations of the company, usually only impacting on certain departments within the organisation.

- *Opportunity*. Not necessarily crucial to the current operations of the company, but seen as offering potential future benefits.

No doubt benefits which may be classed as operational for one company are of such a major importance to another company that they must be termed strategic. In this respect the categories are not clear cut.

Strategic

Faster Trading Cycle

This is particularly important where speed and accuracy of the ordering and invoicing systems is vital to main business operations. For example, it is no use if the travel agent receives out-of-date holidays. Figure 4.1 illustrates typical time savings provided by EDI trading when compared to use of the postal system.

WHAT DOES EDI PROVIDE?

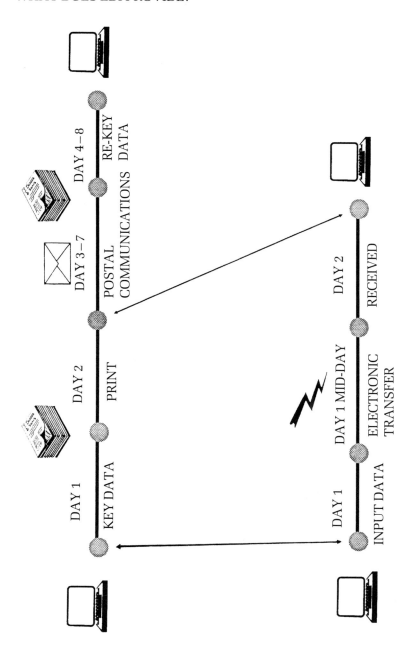

Figure 4.1 Typical Time Factors Associated with Electronic and Manual Trading Systems

14   WHAT IS EDI?

Just-in-time Manufacturing

In essence, just-in-time (JIT) manufacture refers to the ability to produce minimal-sized batches of finished goods only when needed, ie responding to the market 'pull'. The JIT philosophy encompasses zero defect quality, improved communications and logistics, and flexibility and versatility in both the workforce and the machinery. A hallmark of JIT manufacturing is JIT purchasing, whereby small quantities of raw material stock are ordered at frequent intervals, or just-in-time. Without JIT techniques, high levels of raw material, work-in-progress, and finished goods stock will need to be maintained to cater for the slow response to such factors as unpredictable customer demand, shortage of materials, and research and development imposed changes to production processes. EDI, by speeding up the sales and purchase trading cycle, facilitates the adoption of these JIT techniques.

Terms of Trade Dictated by Bargaining Power

In a highly competitive market EDI can be a powerful weapon in a company's corporate armoury. The ability to offer significant cuts in product delivery time may be the difference between winning and losing a contract. This is particularly relevant to international trade, where the movement of paperwork can lag behind the associated movement of goods, resulting in the goods being tied up in port waiting for clearance. The net effect is increased cost to the carrier, which filters through into increased cost to the customer, and a likely loss of market share.

Need to Respond to Highly Competitive Market Entrants

It is often stated by those involved in promoting EDI that 'if your company does not adopt EDI, you can be sure that others will'. It is not difficult to point to dinosaurs within industries, dinosaurs which have failed to adapt and evolve, and been driven into extinction by highly competitive market predators.

Operational

Reduced Costs

This has long been seen as one of the major savings offered by EDI, and includes the following cost reductions:

WHAT DOES EDI PROVIDE?

15

- Paper and postage bills cut.
- Reduced stock holding costs: by speeding up the ordering cycle and providing better information on product sales, EDI enables minimum stocks to be held by the manufacturing process. The consequent cost savings in floor space and storage, insurance, and the reduced need for security and warehousing staff may be very significant.
- Manual processing costs: these include the verification, keying and re-keying of documents, and the cost of manual filing systems. This cost will become more apparent as manual administrative costs increase whilst computer system costs decrease. It is worth noting that the receiver of data (the customer) will normally achieve the higher benefit, because of the impact of computing technology on manual data processing methods. Indeed, in the ECIF/AFDEC trial (see Chapter 9) within the electronic components industry, Mike Pickett of Philips Components (formerly Mullard Ltd) has reported "massive savings on incoming documents, where the cost of receiving information was only 20% of the paper equivalent costs. The cost per 1000 documents, including mail handling, computer cost, verification/ entry and error handling amounted to £1510 using paper-based methods, which fell to £325 using EDI, a saving of nearly £1200."

Improved Cashflow

The importance of cashflow to business requires no elaboration. EDI enables suppliers to send accurate and timely invoices. Customers report vast improvements in matching invoices with purchase orders and deliveries, resulting in reduced errors and queries. The result is that a higher proportion of invoices get paid on time.

Security and Error Reduction

This includes the elimination of errors in transcribing documents from one medium (paper) to another (computer), and the reduced chance of mismatching orders, ie fewer goods returns, and one delivery of the correct goods.

Acknowledged Receipt

A feature of many EDI systems is an automatic confirmation of delivery of the electronic document.

Opportunity

The list of opportunity benefits will include such factors as enhanced image and competitive edge, which although perceived as beneficial, are difficult to quantify. The appearance of screens on travel agents' counters, whereby a holiday may be booked electronically is one such example. These factors give rise to new business opportunities, arising from the better service given to trading partners. As many companies begin to insist on EDI trading with their suppliers, so the company offering this service will enhance its chances of securing a wider choice of trading partners.

THE IMPACT ON ORGANISATIONS

It is predicted that EDI will revolutionise the way companies do business. This is a bold statement to make for what is, in the end, just another computer-based system. Certainly, such sweeping statements have not been made for payroll, order processing or other computer-based systems. One crucial difference between these systems is that, unlike in-house developments, EDI is a cooperative system, shared between trading partners. This difference underlies many of the reasons for EDI's dramatic impact on business. In order to gain some measure of this impact, it is useful to examine the differences between the development of an EDI system and typical in-house developments.

Greater Cooperation Between Customers and Suppliers

This essential cooperation results in a greater understanding of the operation of each other's business needs and associated problems, ie a better working relationship.

Standards

These play a far greater role in EDI than for in-house systems. The lack of standardisation between computer manufacturers has often been loudly stated by the user community as a major hindrance to the widespread adoption of information technology. Companies cannot afford to develop IT strategies in isolation from their existing range of hardware. This is particularly true of EDI, where it is important for organisations to be able to 're-use' the EDI system for many trading partners. Not only might each partner be 'allied' to a

WHAT DOES EDI PROVIDE? 17

different manufacturer, but they are likely to use different forms of the same document (the electronic equivalent of speaking a different language). This factor has led to the requirement for industry, national and international standards.

Importance of Education of Trading Partners

For an in-house development user education is relatively simple, since it is assumed that all employees are working towards a common business strategy, and the 'rule of law' can always be enforced from senior management. These characteristics are unlikely to apply to the EDI system, where the two businesses may have very different and conflicting goals. The promotion of EDI must be approached with much more care.

Involvement of Third Parties

Although true of in-house systems development, where a company may employ the services of external consultants, the major difference for EDI is in the development, maintenance and promotion of standards, and the provision of communications links between trading partners. These two factors are vitally important in helping to reduce the complexity of interconnecting, transmitting and interpreting data, where multiple trading partners are involved. They are explored in detail in Chapters 6 and 7.

Methods of Working May Need to be Re-evaluated

It is true of all computer-based systems that, when analysing the requirements for these systems, old methods of working should be examined with a view to rationalisation and simplification. There is no law which states that the computer-based system should mimic the paper-based manual system it replaces. In EDI terms, this may imply that the invoice accompanying a shipping notice, for example, could prove to be redundant. Suppliers to the motor industry are finding that EDI is effective in supporting the new practice of self-billing. Here the vehicle manufacturer raises an invoice for goods received and sends this back to the supplier for reconciliation against company data.

With working practices, the importance of this examination stems from the effect that EDI has on the timing of the business trading cycle. This will undoubtedly be altered, with a dramatic reduction in the waiting time involved in the receipt of documentation. Companies will need to adapt to this change in timing.

Technical Aspects Are Not the Major Issue

The major challenge in establishing EDI trading is not a technical one; the development and installation of EDI hardware and software can often take place within several weeks. The EDI challenge is one of effecting a cultural change within organisations, building new 'electronic relationships' between trading partners.

Need for Openness and Lack of Secrecy

Organisations implementing EDI trading cannot afford to adopt the air of secrecy which often characterises in-house computing developments. Promotion of EDI is very important, particularly where new standards to the content and presentation of documents are involved. Industry partners must be encouraged to adopt such standards by a vigorous marketing of EDI benefits.

ASSESSMENT OF RISK

There is no doubting that the implementation of EDI offers significant benefits to company trading. However, any sound business strategy must also address itself to the risks involved in this implementation plan. These 'negative factors' are given below.

Lack of Interconnection/Interworking Between Communication Networks

This is exemplified by the motor industry, where two third-party networks, EDICT and MOTORNET, together with Ford's internal FORDNET network, are in widespread use for EDI trading, the first two being endorsed by the Society of Motor Manufacturers and Traders. Traders may be obliged to establish links to all three networks, with the consequent costs.

The technical aspects of interconnection present few problems but such issues as 'who pays for what?', 'where does responsibility lie?', and the thorny question of loss of trading revenue to rival network suppliers must be addressed. The network suppliers, under

WHAT DOES EDI PROVIDE? 19

pressure from the user community, are beginning to respond. Communications networks are explored in detail in Chapter 6.

Legal Aspects

In common with many rapidly developing technologies, the law has been slow to respond to a changing world. In the absence of a secure legal framework, it has been left to bodies such as the International Chamber of Commerce (ICC) to develop guidelines for conducting EDI. The legal aspects include the provision of new forms of contract, and deciding when and where the contract is formed, an issue of some importance where EDI transactions cross national boundaries. Possible computer fraud and theft of data (can electronic data be stolen?) need also to be addressed. These legal and security issues are examined in Chapter 11.

Negotiable Documents

The Bill of Lading is used in the shipping industry as a document giving title to the goods in transit, and must be produced (properly endorsed), before the purchaser can take possession of the goods. This so-called negotiable document has proved difficult to replace or transfer onto an EDI-based trading system. The alternative Waybill, which is appropriate where payment for the goods is assured or has already been made, lends itself to EDI and is being heavily promoted.

Limited Number of Suitable EDI Packages

EDI software is necessary in order to interface between user applications and the outside telecommunications world. Up to now, this requirement has been largely addressed by the establishment of utility packages for easing the conversion of in-house document standards to EDI equivalents, and by the development of EDI workstations handling document selection, validation, batching, addressing and audit. However, there are still few application packages providing turnkey solutions to specific business functions. Chapter 6 describes the requirements for these enabling products.

Limited Number of Standard EDI Documents Defined

The development of EDI standards is crucial to the widespread

20   WHAT IS EDI?

adoption of this form of trading. The foundations to a truly international set of document standards have been laid in the form of the EDIFACT syntax standard. However, much work still needs to be done to establish a wide range of standard documents that conform to this syntax standard. The role of standards is explored in Chapter 7.

Critical Mass of EDI Users not yet Reached

It is symptomatic of electronic office communications technology that a critical mass of users is required before a 'big bang' style adoption of the techniques is implemented. Potential users must assess whether the population is sufficient for them to realise their investment, and certainly some business sectors are more mature than others in this respect. In the 'green field' situation, where no EDI trading currently exists, the advocate must assess the risk involved in trading partners not taking the EDI bait. Chapter 9 assesses the development of EDI within various business sectors.

Wide Range of Participants to a Business Transaction

It may be a daunting prospect for potential EDI implementors to consider the number of parties involved in a business transaction, particularly so in international trade where the participants include buyer, seller, banks, agents, carriers, customs, insurers and government ministries. Agreement must be reached on standards, communications media and legality. However, as the number of participants increases, so too does the scale of benefits, for all those concerned. It is significant that many of the early and most successful EDI implementations have taken place in the international trading arena (refer to Chapter 9 for a discussion of HM Customs and Excise systems).

Network Collapse

If trading partners rely on third-party networking, what allowance must be made in the unlikely event of a prolonged failure of the network? Rival network suppliers must be assessed for the adequacy of failsafe devices, or even the options available in the (unlikely) event of the network service being completely 'written-off'. Can manual methods be reinstated should the need arise?

WHAT DOES EDI PROVIDE? 21

CONCLUSION

EDI is providing many organisations with a corporate weapon of immense value, as they strive to become more lean and effective in a volatile marketplace. The EDI business strategy requires not only a realistic assessment of benefits, but must seek to quantify potential risks, and appreciate the likely impact on the organisation. Suppliers and consultancy organisations are only too eager to offer skills to aid the wary.

In concluding the overview, Chapter 5 provides a sample scenario of EDI in practice.

5 What Does EDI Look Like? – Sample Scenario

As the reader may have gathered, EDI is implemented for many reasons and consequently takes many forms, depending on the business sector and commercial requirements. The following fictitious example is not intended to cover the spectrum of possible implementations, but rather to present one picture (out of very many) of EDI 'in action'.

In this example the trading partners are both manufacturing companies, 'Widget Brothers' supplying components to its dominant customer 'Megatrader Incorporated'. The scenario is typical of many trading relationships in the motor, electrical and retail industries. Prior to EDI, the future of the two companies looks less than rosy.

IN THE BEGINNING ...

Widget is under pressure from Megatrader for an improved level of service. Megatrader has progressively enforced a reduction in the lead-time for fulfilling orders (they mostly arrive on Thursday and require to be despatched by the following Monday), and is insistent on a very high standard of quality. Its wide product range has meant that Widget finds it extremely difficult to predict the orders in advance. Widget has been left with no option but to adopt a costly weekend production run and to tie money up in maintaining high stock levels. All of this is placing a great strain on cashflow, exacerbated by the fact that Megatrader are less than prompt in making payments. Staff morale is at a low level, Tuesday afternoons being particularly troublesome with phone calls about missed or inaccurate deliveries.

From Megatrader's point of view the situation appears to be

24 WHAT IS EDI?

equally bleak. The market for its products is very competitive and subject to fluctuations. A recently-opened Japanese plant in particular has gained a reputation for quick delivery of high quality products, forcing Megatrader to reduce costs where possible, vary the production run, and minimise defects. Megatrader are well aware of the problem that suppliers are having, particularly with cashflow, and would like to help. Unfortunately, the clerical staff are having great difficulty in interpreting the many types of invoices received, each with its own layout, and often with inaccuracies, resulting in phone calls back to the supplier, re-invoicing, etc.

THEN CAME EDI ...

It was timely for both companies when they received an invitation to attend an EDI seminar. They immediately saw the value of EDI as a means of improving the flow of information between themselves, and agreed right away to implement a trial. Production schedules and invoices were seen as the two areas crucial to improving service levels.

Megatrader developed in-house software for converting their production schedule data formats, held on minicomputer, to the recognised and widely adopted industry standard. These standard production schedules are 'posted' electronically by the minicomputer to an electronic postbox, provided by a network supplier offering 'value added' services. These services were seen as particularly important to both Megatrader and Widget, since they include comprehensive network management, security and auditability in addition to the basic 'mailing' facility. Once the Megatrader computer has deposited the schedules in its postbox, its work is complete. The network provider 'reads the address' on the schedules and transfers them to Widget Brothers' electronic postbox. Widget Brothers, at their convenience, then transfer these schedules down to their own personal computer. The personal computer approach, utilising tried and tested packaged software for converting to their own document formats, has provided Widget with a simple, low cost method of implementing and evaluating EDI.

Megatrader have their own electronic network postbox which they empty daily, and use to receive Widget's invoices. These invoices are then transferred automatically into their computerised accounting system.

WHAT DOES EDI LOOK LIKE? – SAMPLE SCENARIO 25

AND NOW ...

The trading situation now looks much different.

Widget Brothers now receive Megatrader's production schedules minutes after they are produced on Tuesday and can gear up production so as to despatch by Friday. Weekend working has disappeared and Widget have themselves moved towards manufacturing 'just-in-time', rather than the previous system of 'always-behind-time'. Megatrader has now disappeared from the debtor list, due mainly to the transmission of accurate EDI invoices. Staff morale has improved one hundred percent.

Megatrader is highly delighted with the competitive edge that EDI has provided it with. Being able to order parts 'just-in-time' is seen as an essential strand in the new flexible, high quality manufacturing environment. The company has now started to analyse other areas of its operations which could benefit from EDI. In particular it has already commenced discussions with transporters, shippers, freight forwarders and customs to speed up the flow of exports.

EDI has given both companies the breathing space and information to anticipate and pre-empt problems. They agree that administrative costs (postage, phone bills, etc) have no doubt decreased, although by how much they do not know, since neither company has yet worked this out. The friction that previously existed between the two companies has been replaced by a much closer understanding of each other's problems and requirements. Widget Brothers are particularly pleased, since they have been assured by Megatrader of 'Number 1 supplier status'. They have also won many supply contracts that previously they had been losing, since now they are gaining a reputation for a high level of service.

BUT, BACK TO REALITY ...

'Widget Brothers' and 'Megatrader Incorporated' do not exist, although the trading scenario may ring true to many readers. In the realms of fiction it is easy to gloss over such problem areas as the establishment of communications lines, ensuring that the hardware and software is correctly configured, not to mention the cost of installing such a system. However, the method described by which EDI operates is now commonplace, and the associated benefits are typical of those realised by companies in the manufacturing sector.

Megatrader and Widget would surely agree that EDI would have been difficult to establish but for the availability of secure and well-managed EDI communications networks, together with accepted standards for transmitting and presenting the data. These two aspects of EDI are now discussed in Chapters 6 and 7.

6 Communications Issues

It is significant that the first strategic benefit of EDI to be listed in Chapter 4 is that of speeding up the trading cycle, since from this stem many of the other benefits, such as the implementation of just-in-time ordering. Central to the realisation of this benefit is the establishment of an efficient and effective communications infrastructure. This may be viewed as the 'EDI plumbing', whereby trade data enters the (sender's) inlet pipe, and flows through the system to emerge at the (recipient's) outlet pipe. In addition to the communications 'pipework', the EDI system requires a very high degree of reliability and availability, an excellent support service, alternative 'pipelines' in the event of loss of the usual service, and the ability, if necessary, to trace the path of a trade document through the system.

This chapter considers the alternative methods of satisfying the above requirements, and the reasons why many companies are choosing to use the services of third-party vendors.

TRANSMISSION MEDIA

Magnetic Tape

A simple and cost-effective means of passing data from one computer system to another is by transferring, or 'off-loading', that data onto magnetic tape. The data is stored electronically on the tape, in much the same manner as an audio cassette tape is used to hold voice data. The tape is then transferred manually, either by postal or courier service, to the receiving computer installation, and 'loaded up' onto that computer system. Indeed this is the accepted method used by software suppliers to install both new software products and updates to existing products at remote customer sites. In this case, speed of delivery is not usually crucial; however, it is essential that the data is transferred error-free, and in a manner in which it can be easily handled by the existing software applications and tools. These same issues are present where magnetic tape is used as the EDI transfer mechanism, and the corresponding benefits ensue. Consequently, although this transfer method does produce significant time-savings in the elimination of the error-prone data keying operations, it may not produce the time savings necessary to dramatically alter the trading cycle. The method is most suitable to the transfer of large batches of data, which may offset the cost of producing and delivering the tape. The ease of implementation probably explains why this method is in widespread use.

The alternative to magnetic tape is a telecommunications network, where the line medium may be physical (wire, optic fibre) or non-physical (cellular radio, satellite). These telecommunications options are now discussed.

Public Switched Network Options

Switched networks allow an organisation to communicate with many other companies without having to establish dedicated links to any of them. The network may be either the Public Switched Telephone Network (PSTN) or a packet switched service such as BT's PSS in the UK. The choice would depend on such factors as the need to support higher level data and file transfer protocols.

Companies choosing to subscribe to the PSTN would require a modem in order to convert digital signals to audio frequency, suitable for transmission via the analogue telephone network. The method of accessing the packet switched network depends on the size of the computer system. For example, for a small personal computer this may involve little more than the installation of a communications card plus network connection software.

Private Networks

Where the amount of data to be exchanged between organisations is particularly high and/or at frequent intervals, companies may wish to consider leasing a private circuit, from either BT or Mercury in the UK. This allows for a dedicated (point-to-point) link between trading

COMMUNICATIONS ISSUES

partners. This option becomes unmanageable and expensive when a company requires to communicate with many organisations on an infrequent basis.

The private circuit may allow direct digital transmission, in which case access is via a simple Network Terminating Unit (NTU).

Whether the network is public or private the trading partners will need to agree on a set of rules governing data transmission, including:

- speed of transmission;
- data encoding format (ASCII or EBCDIC);
- synchronous or asynchronous transmission;
- communications protocol;
- time schedules for transmission.

This last factor may well be the most serious communications problem to be overcome for a small company that does not keep an 'open door' to receive EDI transmissions. In this case the trading partners are required to agree scheduled times for transmitting batches of data. This data can be transferred when, and only when, the two computers are ready to send and receive respectively. This may prove extremely difficult where international data transfers, involving different time zones, are involved.

The need to satisfy the above problems has caused many EDI implementors to consider the services of third-party network vendors, offering a range of Value Added and Data Services (VADS).

VADS

What are VADS?

As previously stated, VADS stands for Value Added and Data Services. In simple terms they involve two components:

- a network; and
- an application.

The application provides the added value to the basic data networks

30 WHAT IS EDI?

already described, allowing users to purchase such facilities as electronic mail, on-line database access, and Electronic Data Interchange. In the UK, the network will be provided by either British Telecom or Mercury, currently the only two companies licensed to own and operate a fixed-link network which can be sold to third parties. The third-party VADS suppliers lease transmission capacity from either of these companies, to which they then add value using their own computer processing capacity. (VADS were formerly called VANS, Value Added Network Services.)

What Services are Provided for the EDI Community?

The following services are typical of those provided by VADS suppliers.

Network Service

At the basic level, this implies a comprehensive electronic interconnection service between trading partners, with high levels of security and system resilience. Typically, the network may provide multiple routing paths and physical links. Network management will be provided to ensure both the day-to-day control and the smooth implementation of any changes. Users will need to be assured of maximum availability and performance, taking into account the requirements of the supplier to carry out the necessary network maintenance.

Postbox/Mailbox Service

This service is often referred to as the central 'clearing house' or 'store-and-forward' function of the third-party VADS network. Figure 6.1 illustrates the essential features. Each VADS user, or EDI trading partner, is provided with an electronic postbox or mailbox (in reality an area of storage within the vendor's computer system). The sender of the EDI transmission deposits the data, suitably 'packaged' with the recipient addressing information, into his postbox. The data is stored here, prior to processing, when the 'address is read' and the data forwarded to the recipient's mailbox, for subsequent collection.

The main benefit of the postbox/mailbox service to the EDI user is that it frees both the sending and receiving computer applications

COMMUNICATIONS ISSUES

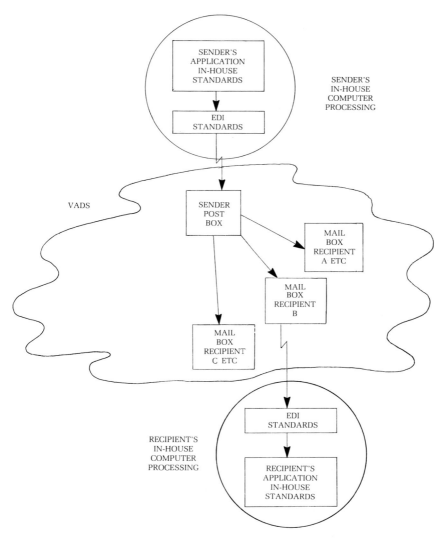

Figure 6.1   The VADS Approach to EDI

from the requirement to synchronise data transfer, one of the main problems with the direct link method. In addition to this basic mailbox service, suppliers may offer an archive service, on-line mailbox processing (view and delete), automatic confirmation of delivery, and priority mailing. Users must decide if these features are useful, since they may be charged extra for them.

Enabling Software Service

The VADS supplier will often also provide software both for connecting to the network and for converting in-house document standards to and from the adopted trade standard. This service may be provided by a separate software company. These enabling services are considered at the end of this chapter.

Consultancy Service

The consultancy may cover a needs assessment (the 'business analysis'), advice on standards, and a comprehensive implementation package.

Support Service

The widespread adoption of EDI trading relies heavily on strenuous promotion in order to attract additional trading partners. As one might expect, since the VADS supplier has a vested interest in attracting additional subscribers, ie paying customers, this service receives vigorous support. The support services may also include such activities as executive seminars, user training, hotline support, and the representation of business needs to trade associations.

What are the Benefits?

The following list of benefits are typical:

- An 'on-demand' service, 24 hours-a-day.

- Wide geographical coverage, with local access to the network from many sites and locations.

- Support for most major communications protocols. Each trading partner requires only one communications link to the VADS bureau. The sender and receiver may use different communications protocols.

- The bureau will support all the widely used document and data formats, thus allowing inter-industry document interchange. These formats are discussed in Chapter 7.

COMMUNICATIONS ISSUES 33

- The store and call forward mailbox facility allows the user total control of the scheduling of outgoing and incoming transmissions.

- Small companies may enjoy the benefits of EDI as easily as large ones. The low cost and flexible packages and workstations provide the small trader with a cost-effective EDI solution.

- Viability for low volumes of trading transactions. This is particularly important where an organisation's trading is not dominated by a few large revenue customers, but consists of many small revenue trading partners. In this situation a one-to-one direct link *for each trading partner* would not be appropriate.

Main VADS Suppliers

In the UK, the two leading VADs suppliers offering EDI are INS and Istel. The vendor market includes IBM and DEC. Each of these suppliers has approached the provision of EDI VADS services from a different stance, influenced by the existing network and customer base and support from trade associations.

IBM UK

EDI service: This is based on the IBM Information Exchange service.
Business sectors: These include transport, shipping, the London insurance market (LIMNET), construction, retail and distribution, manufacturing, and government (of Singapore).
Network: IBM's international Managed Network Service (MNS).

INS (International Network Services)

INS is a joint venture company set up by ICL and GEISCO, in January 1987.

EDI service: INS has 'brand-named' its EDI service to the business sector, and offers the following:

- TRADANET

   Using the TRADACOMS trading standards (see Chapter 7), this is the most widely used EDI service in the UK. It has mainly been aimed at the consumer goods industry, and includes (corporate) manufacturers, wholesalers, distributors and retailers.

34 WHAT IS EDI?

— TRADANET International

Aimed at the international trade market, its users include the shipping, export and freight industries, banks and insurance.

— DRUGNET

Brand-named for the health care industry, its users include doctors, pharmaceutical companies and government departments.

— MOTORNET

The INS EDI service for the automotive industry.

— BROKERNET

For insurance companies, brokers and Lloyd's syndicates.

— PHARMANET

The EDI service for the pharmaceutical industry.

Network: INS use the Mercury network in the UK and GEISCO's international network outside the UK and in over 70 countries.

Istel

Istel is a leading UK systems house, based in Redditch.
EDI service: EDICT.
Business sectors: EDICT serves users in the manufacturing, travel, health, finance, retail and distribution, and automotive sectors.
Network: EDICT runs on Istel's Infotrac carrier network.

DEC

DEC are a late entrant into the VADS market, providing services for the financial sector (brokers, solicitors and estate agents), covering mortgages and life insurance.

Other VADS Suppliers

It is worth noting two other VADS suppliers. Travinet is a

COMMUNICATIONS ISSUES

subsidiary company of Thomas Cook, itself a subsidiary of Midland Bank, and offers a videotex service, running on its national Fastrak X.25 packet switching network. Although Travinet has not yet penetrated the EDI market, the scope of its network, rivalling BT's Packet Switchstream (PSS) service, makes it a potential candidate for the future.

British Telecom itself has tested the EDI waters, and provided what is generally considered to be the first implementation of EDI in the (now obsolete) Heathrow airport LACES system. Following on from this, BT joined forces with McDonnell Douglas to launch Edinet, an EDI service targeted at the retail and distribution trade. Although the venture collapsed at the end of 1986 (due to a failure to secure user support!) both parties are widely tipped to re-enter the EDI arena at some future date.

How Much Does it Cost?

In assessing VADs suppliers, potential subscribers will wish to consider the existing and likely future customer base, in addition to the charging structure. In general, the following costs will have to be taken into account:

- registration charge
- dial-up charge
- leased line charge, based on number of characters
- network line charge

Also, users are likely to be charged for the amount of data stored in both the mailbox and archive storage. The charging philosophy adopted by the supplier should be considered, ie does only the sender pay, or do both the sender and the recipient pay?

THE ENABLING SOFTWARE

The most sophisticated network services are to no avail if there is no adequate means of linking existing in-house applications to these networks and of making sense of the EDI data received from the network. Many companies, particularly those with good in-house data processing resources, may wish to develop their own software for this task. However, the availability of reliable, user friendly and

36

WHAT IS EDI?

well-documented packaged software seems essential if the growth of EDI is to be maintained, particularly so for the small trader.

The EDI packaged software market is now beginning to flourish with products available for micro, mini and mainframe systems. The most established of these is SITPRO's Interbridge software, which provides a hardware independent solution to the formatting and deformatting of data for electronic data interchange. It implements the UN/TDI (eg TRADACOMS and ODETTE) and EDIFACT syntax rules (see Chapter 7). Interbridge is an application independent product which may be used to link to existing user applications or to form the basis of a self-contained EDI workstation.

IBM have enhanced the functions of Interbridge to provide control and audit functions, data validation, and batching and addressing. This additional function is provided by their EDILINK package.

Several packages which have been developed for the exporter are worthy of mention. EQUATOR, from INS, runs on the personal computer and provides facilities for entering information, creating EDI messages and interfacing to other systems. It has been designed for the international trade market, linking exporters to other exporters, importers, carriers, freight forwarders, etc. SPEX from SITPRO, is not, strictly speaking, an EDI package but it allows the exporter to store and process product and customer information. It thus speeds up the process of producing invoices and shipping documentation, and may be used as the user-interface to an EDI trading system.

In addition to the network vendors, software houses have also been active in the supply of EDI software. The company Systems Designers have developed an EDI workstation to handle the creation, sending and receiving of messages. The workstation is a general purpose product, running in either the micro, mini or mainframe environments, which may be linked by user-specific software to in-house applications.

The above discussion is intended to give the flavour of what is presently on offer in the enabling software field. As competition develops one can expect the product range to improve and the costs to reduce. Users should bear in mind that many of these packages are flexible, general purpose products that require the careful design of

COMMUNICATIONS ISSUES

37

friendly, user-specific 'front-end' applications for successful implementation. The need to invest in consultancy and training should not be ignored.

CONCLUSION

In covering the communications issues, this chapter has dealt in some detail with the VADS option. This may be justified by virtue of the benefits that VADS provides for the EDI community, benefits which will increase as more trading partners are introduced to the service. The UK government has recognised the importance of VADS through the launch in October 1986 of Vanguard, a joint initiative with industry to promote the increased use of Value Added and Data Services. Vanguard has been a catalyst in the growth of VADS. At the time of writing the future of Vanguard is uncertain; however, the future of VADS to the EDI community seems assured.

7 Standards

There can be few publications devoted to the technology of computing which do not at some stage refer to the bugbear of standards. This may involve consideration of the problems of connecting two different (or even the same!) manufacturers' hardware together, the problem of different software packages 'talking to one another', or the difficulty in maintaining software that has been developed without respect for any design and programming ground rules. These problems will be familiar to most organisations developing, maintaining and enhancing in-house computer systems. However, the analyst developing an in-house system can at least specify the format of data input to the system and reports produced by the system. The EDI analyst is faced with catering, for example, with every type of purchase order received by the company, not only from existing customers, but from potential customers in the future. It is not surprising therefore, that the role of standards has played such an important part in the development of EDI and continues to do so.

This chapter will consider the development of industry national, and international standards, and the options available for the newcomer to EDI.

WHAT ARE EDI STANDARDS?

EDI standards refer to standardised ways of describing data items (article number, article price, unit price, name, street, postal code, etc), and of grouping and presenting these data items in the form of messages or trade information (invoice, purchase order, etc). EDI standards are therefore not communications protocols, and not part of the Open Systems Interconnection (OSI) seven layer model for data communications, although it may be helpful to view them as perhaps the eighth layer of this model.

WHY ARE EDI STANDARDS IMPORTANT?

This may be illustrated by considering the situation in which two companies agree to adopt EDI for the transfer of each other's purchase orders. In this simple case two (in-house) standards are involved, and each company requires software for converting the other's in-house standard to its own. If four companies are now involved in this form of EDI trading, clearly four (in-house) standards are involved, and each company must now acquire software to convert three other standards to its own, implying 12 different conversion routines (refer to Figure 7.1). Clearly this situation is neither sensible nor efficient, and the involvement of more and more trading partners will only make matter worse.

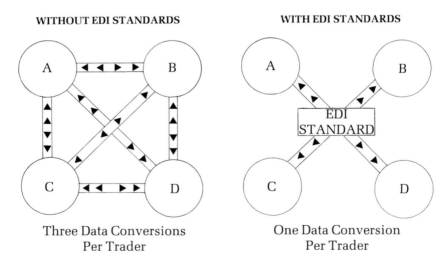

Figure 7.1 The Complexity of Converting Data to In-house Standards, Illustrated for Four Trading Partners

The desirable solution is for all trading partners to agree to adopt a common standard for their in-house systems, which would remove any requirement for conversion routines. This would, however,

STANDARDS

41

necessitate considerable change to these systems, and would reduce the likelihood of attracting new users to the fold! The alternative and practical solution is for a common EDI standard to be adopted when transferring data. In this case each company requires only one conversion routine no matter how many trading partners are using EDI.

WHAT FORM DO EDI STANDARDS TAKE?

EDI standards have been structured with respect to a reference model, consisting of the following components:

- *Data Element*. Data elements provide the vocabulary of EDI, and identify those individual fields or items of data designed for a specific purpose, eg product code, expiry date, unit price, invoice number, postal code. Individual data elements may be combined to form composite data elements, eg a weight of 24 kilos is represented by the composite data element 24:KG. The 'dictionary' that holds all these data elements together is referred to as the data element directory.

- *Segment*. A segment is a functionally related group of data elements or composite data elements, eg supplier name, address and reference ID.

- *Message*. A message is a group of segments brought together for a specific purpose and sent electronically. Examples of messages include the invoice, purchase order, credit note and waybill. Messages are required to adhere to message design guidelines and the syntax rules (see below).

- *Functional Groups*. These refer to groups of messages of the same type, eg all purchase orders to one company.

The EDI reference model provides the 'building blocks' for constructing EDI standards. Figure 7.2 illustrates how the model relates to a set of company invoices. The standards developer also requires guidelines or rules which state how the building blocks should be assembled.

Syntax rules state how individual data elements or composite data elements should be grouped together, and message design guidelines allow groups engaged in designing new messages or modifying

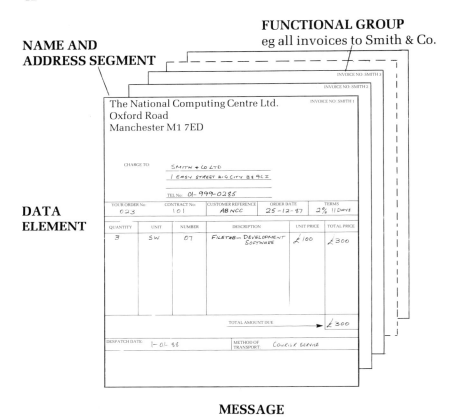

Figure 7.2 EDI Standards Terminology Applied to a Collection of Company Invoices

existing messages to do so in a consistent manner, which will allow other users to understand them.

HOW HAVE EDI STANDARDS DEVELOPED?

The development of EDI standards mirrors the history and development of EDI trading itself, as one might expect.

In the UK, as early as the 1970s, the Simpler Trade Procedures Board (SITPRO), a government-sponsored body, produced a set of

STANDARDS 43

syntax rules and a dictionary of data elements for trade data exchange. SITPRO's work was adopted by the United Nations Economic Commission for Europe. The UNECE produced a Trade Data Element Directory (UN/TDED), which has since progressed to full international standard status (ISO 7372). In addition, they issued a standard syntax for the data interchange in the form of guidelines for Trade Data Interchange UN/TDI. These guidelines have been used as the basis for data exchange by many industries and applications, mainly in Europe. Trade associations, keen to promote the use of EDI, have adapted the UN/TDI syntax standards to produce message standards for specific business sectors, eg TRADA-COMS for the retail trade, ODETTE for the motor industry. These message standards are considered in more detail in a later section.

In the United States interest focused on the work of the American National Standards Institute (ANSI) X 12 committee, culminating in the release of standards for:

— transaction sets (messages);

— a data element directory;

— transmission control standards (syntax).

ANSI X12 has been widely adopted throughout America, but has failed to make significant inroads elsewhere (though there is some ANSI X12 use in Australia).

This resulted in separate standards for Europe and America. In an attempt to resolve this situation the UN-JEDI group (UN joint EDI group) was formed, with members drawn from the United States and Europe (both East and West). The group was charged with drawing together the ANSI X12 and UNECE work, the first task being to produce appropriate syntax rules. The work so far may be summarised as follows:

— The Trade Data Element Directory (TDED).

— The Syntax Rules. This work is now complete and the results have been issued as an international standad ISO 9735 called EDIFACT (Electronic Data Interchange for Administration, Commerce and Transport).

44   WHAT IS EDI?

- An implementation guide to EDIFACT.

- EDIFACT Message Design Guidelines.

- EDIFACT Standard Messages. The first one to be defined is the Specification for UNECE Standard Electronic Commercial Invoice Message.

EDIFACT provides the basis for a truly international EDI standard, and is being widely promoted by EDI 'champions' such as SITPRO. New EDI implementations are already adopting EDIFACT (eg the CEFIC project within the chemical industry, see Chapter 9). However, well established EDI projects, adhering to TDI-derived industry standards, may not be so easily convinced of the benefits of moving to the EDIFACT camp. Two of these industry standards are discussed below.

TRADACOMS

The TRADACOMS standard originated from work done by the Article Number Association (ANA), an independent trade grouping funded primarily by the consumer goods industry. In 1981 the ANA set to work on a set of standards for all documents to be exchanged between trading partners. The result was TRADACOMS, a comprehensive set of EDI standards, covering invoices, price lists, orders, delivery notes, credit notes, etc.

In 1983 the ANA issued a tender to third-party VADS bureaux, to provide a network supporting TRADACOMS, the endorsement eventually going to ICL and its TRADANET service (TRADANET is now supplied by INS). TRADACOMS is now the most widely used EDI standard in the UK, and is supported by all the major VADS suppliers. Note that although TRADACOMS uses the UN/TDI syntax rules, it uses its own trade data element dictionary and not UN/TDED.

ODETTE

ODETTE, the Organisation for Data Exchange by Tele Transmission in Europe, was formed in 1985, by motor manufacturing companies and their suppliers in eight European countries. ODETTE has concentrated on developing standard commercial messages (such as

STANDARDS

invoices and schedules) for EDI interchange, all based currently on the UN/TDI syntax standards and using the UN/TDED data dictionary. ODETTE is now committed to changing to the EDIFACT syntax.

Figure 7.3 summarises the development of standards and the relationship with the sponsoring trade associations.

To complete the picture, mention should be made of the many countries excluding the UK and US, who have developed their own national standards. Primary examples include West Germany, Holland, Belgium, South Africa, Sweden, Australia and France, where the national article numbering authorities have created EDI standards systems tailored to their national trade requirements.

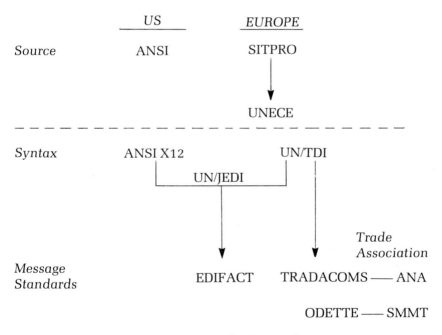

Figure 7.3 EDI Standards Development

WHAT NEXT FOR STANDARDS?

EDIFACT developments will continue, and one can expect to see its use opened up to a wider range of business sectors, eg book trade, tourism and leisure. The development of EDIFACT standard messages is well underway and, in addition to the draft invoice currently available, one can expect to see a purchase order, confirmation of purchase order, and despatch advice in the not-too-distant future. Messages for the transport, customs, banking and payment areas are expected to follow.

Established EDI projects may be expected to begin trials with the EDIFACT standards; already CEFIC have adopted EDIFACT for the chemical industry trial, and the shipping industry EDIS project (formerly DISH within the UK) has announced this commitment for 1988.

The solid foundation for the development of truly international standards is being laid. There is no shortage of vigorous promotion; what is required is the vision and commitment to make it work.

8 Planning the Implementation

It is hoped that the preceding chapters have given the reader sufficient reason and encouragement to believe that the implementation of EDI trading is worthy of active consideration. What next? What steps are involved in implementing EDI? Who should be involved? What decisions need to be taken? This chapter seeks to address these and other related questions. It is impossible in such a small publication to answer all these questions; however, this chapter should serve as a useful checklist for the future EDI implementor.

PRELIMINARY STEPS

Explore EDI

This first step of 'data gathering' implies looking at both the existing trading practices within the company, and the state of EDI in general. In exploring EDI, the analyst will need to address the following questions:

- What is it?

- What are the technical issues?

- What are the potential benefits?

- What are the start-up and operational costs?

- What is the current status within the company?

- What is the current status within the industry? (Trade associations.)

48 WHAT IS EDI?

Involve the Right People

The strategic and functional impacts that EDI can make to an organisation will necessitate that management at both levels is involved throughout the planning and implementation process.

Decide on the Direction

The impact on business practice can be far reaching, as Chapter 4 has illustrated, and will affect:

— Existing computer systems.

— Current working procedures. For example, changes to the timing in the trading cycle, greater accuracy implying fewer follow-up and error checking procedures, and the possible elimination of some procedures altogether.

— Impact on the company. EDI allows new ways of carrying out tasks, as well as making provision for new ways of working. Companies should explore what *should* be done, as opposed to what is being done. Companies may wish to re-structure in order to adapt and take full advantage of the new and more effective methods of operation that EDI enables. These organisational impacts may include:

 • workforce reductions;

 • elimination of certain tasks;

 • task re-specification.

PLANNING TOWARDS EDI

Policy considerations, based on the commercial environment, will include the following:

— Customer base. Whether this is a broad customer base or a small number of customers dominating the order book will influence the choice of network, marketing stance, etc.

— Transaction volumes (per customer per day). If these are low then a higher number of (EDI) trading partners may be required in order to provide a satisfactory pay-off.

PLANNING THE IMPLEMENTATION

49

- Who should pay? Possible algorithms involve dividing the EDI 'running costs' equally amongst trading partners, or each partner to pay for their own transmission costs.

Implementation decisions will need to be taken, covering:

- *Business area and documentation.* The choice of business area and the associated trading documentation (purchasing, sales, accounts, etc). Volume applications, with a high trading paperwork content, should yield the quickest return on investment. In many cases EDI has taken root in the purchasing or receiving departments.

- *Standards.* An assessment will be required of the various options (industry proprietary, national and international) for an EDI transmission standard. Where EDI trading is already well-established within a business sector, this may involve simply rubber stamping and adopting a tried-and-tested standard.

  In a green field situation, the implementation team will need to judge whether existing standards are adequate for handling the required EDI information exchanges, particularly when a choice is made to adopt the new (and yet to mature) EDIFACT international standard. 'Adequacy' here implies assessing whether the message, segment and data elements (see Chapter 7) currently exist and fulfil the requirements. They may need modifying and submitting to the approved standards body.

- *Hardware/Software issues.* Included here will be a decision on where and how to re-format (from EDI to in-house standards and vice versa). Companies may wish to use a dedicated micro or minicomputer to handle reformatting, rather than using corporate machines. Software choices will centre on whether to develop in-house or purchase third-party coding. Allied to this will be the decision on whether to build EDI into existing computer applications (undoubtedly providing the greater long-term benefits, but it is often a major cost), or to adopt the simpler course of manually transferring data between the EDI and existing systems.

- *Network choices.* There are compelling reasons for using the services of a third-party (VADS) supplier (see Chapter 6). Rival vendors will need to be assessed on the scope of the network, facilities and existing customer base.

- *Choice of pilot trading partners.* Financial and operational considerations will obviously be of prime importance here. Questions arising will be 'With which companies do we process the greatest amount of paperwork?' and 'Which of these are both willing and able to undertake EDI with us?'. The effect that EDI has on bringing trading partners 'closer together' will influence the choice in favour of well-established trading relationships, particularly until EDI networks become more widely used.

- *Consultancy.* This should cover an analysis of business needs, leading to a definition of requirements and the technical system specification. These skills may be resourced from specialist EDI consultancy firms, VADS suppliers, or indeed in-house.

- *Training and education.* It is most important that everyone involved in the new trading arrangement should understand it! This implies not only the necessary training in handling new document formats and operating computer workstations, but also an understanding of the company policy that has led to the implementation of EDI trading, especially when considering the education of existing and future trading partners, eg 'Suppliers who join us in EDI will assure their position as one of our prime suppliers'.

WHAT NEXT AFTER THE PILOT?

It has been tacitly assumed that EDI will be implemented via a pilot or trial study, at the end of which the benefits and lessons learnt will be evaluated. The value of this approach is primarily in demonstrating the technology, ie proving that EDI works, and that costing formulae are acceptable to all parties. Following on from the pilot, as well as assigning the EDI service 'production' status, activities will centre on attracting more partners into the EDI club. Here the services of third-party VADS suppliers can be useful in harnessing a natural self interest to increase their customer list. Trading partner conferences are a useful way of promoting EDI growth within specific business sectors, provided they are carefully planned and effectively followed up.

PLANNING THE IMPLEMENTATION 51

As with all pilot studies, EDI practitioners must be prepared to modify EDI procedures in the light of changing circumstances. New trading partners may imply a need for modified standards, and in this respect test procedures that can be reused and updated for each new circumstance will be useful.

9 UK EDI Applications

This chapter provides an overview of UK EDI developments within a wide range of business sectors. This vertical sector analysis provides a useful means of categorising EDI projects and examining the history and growth of EDI within the UK. EDI projects have traditionally formed within closed user group communities, within which standards have been set to serve the member companies. However, since trading relationships frequently cut across these vertical sectors, the future seems certain to see horizontal EDI developments, supported by an interworking of communications networks and a rationalisation of message standards.

AUTOMOBILE

ODETTE

The Organisation for Data Exchange by Tele Transmission in Europe is generally regarded as being the leading European EDI development. ODETTE was formed in 1985 following collaborative efforts by Austin Rover, General Motors, Peugeot Talbot and Ford through the UK motor industry trade association, the Society for Motor Manufacturers and Traders (SMMT). The objective of the project is to facilitate computerised trade data exchanges between automotive companies in Europe. The project currently links vehicle and component manufacturers in eight European countries. ODETTE has concentrated on developing standards for commercial messages (invoices, schedules, etc). ODETTE (UK) endorses two UK VADS clearing houses; MOTORNET, an EDI service supplied by INS, and EDICT, supplied by Istel. In addition, ODETTE endorses various networks in other parts of Europe.

54 WHAT IS EDI?

The problems caused by some companies needing to subscribe to both networks has led to the SMMT pressing for a bridge between these two endorsed services, and a joint project is underway to establish a working link by 1988.

FORDNET

A private network within Ford used to link to its suppliers by PSTN, usually from an IBM PC. Private Ford message standards are used.

MERCATOR

The MERCATOR pilot project involved trading information passed between motor manufacturers, banks, transporters and national customs within the UK, West Germany and Belgium.

CONSUMER GOODS

Many high street retailers exchange data with their suppliers, using the TRADACOMS message standards developed specifically for this business sector. For some of these, for instance B&Q, Marks & Spencer, WH Smith and Tesco, EDI is becoming a routine feature of their supplier trading relationships. The consumer goods sector is mainly served by two network vendors: Istel through its EDICT service, and INS through TRADANET. Using EDI to exchange purchase orders, invoices and stock details allows traders to react to the rapid changes in product lines (characteristic of the consumer goods industry), with minimum costs.

PORTS

FCP80

The FCP80 system was introduced initially at Felixstowe, following the success of the equivalent ACP80 airport system, to improve customs clearance times. Customs data is supplied via Direct Trader Input (DTI). FCP80 was subsequently extended to Harwich, Ipswich, some of the East Anglian ports and to the Medway. At present it is used by nearly 500 subscribers, including port authorities, ship owners and agents, freight forwarders, truckers and government agencies. In addition to DTI, FCP80 also includes an Inventory Control system, recording the identity and status of imports and exports.

UK EDI APPLICATIONS

Similar systems to FCP80 have been installed at a number of UK ports.

AIRPORTS

LACES

LACES is generally regarded as being the first practical application of EDI in the UK. The system was designed to speed up the clearance of freight by customs at Heathrow Airport, and involved the airlines, HM Customs and the forwarding agents. LACES became obsolete in 1981.

ACP80, UKAS, ACP90

ACP80 was introduced in 1981 as the replacement for LACES, and significantly enhanced the airport clearance process. The system began to reach the end of its useful life in 1986 and was replaced by UKAS. However, after only a few days of operation UKAS proved insufficient to meet capacity, and had to be abandoned. ACP80 was eventually replaced by ACP90, when in July 1987 the new system was implemented at Heathrow, Gatwick and Manchester airports.

HM CUSTOMS AND EXCISE

The central role played by HM Customs and Excise in the importing and exporting functions has led.to a significant involvement in both the port and airport EDI developments. The main thrust of developments in the customs arena has been in the use of Direct Trader Input (DTI) and Period Entry (see Glossary) to automate the transfer of information between traders and customs.

DEPS

The Departmental Entry Processing System (DEPS) was launched in 1981. Run by BTAT (British Telecom Applied Technology), the system allows for Direct Trader Input, dramatically reducing cargo clearance times. DEPS is due to be replaced by CHIEF in the early 1990s.

56 WHAT IS EDI?

CHIEF

By the middle of 1988 approximately 80% of import entry information will be passed electronically to customs, a figure which is expected to rise to 90% by mid-1989. In order to cope with this rising demand and to take advantage of the latest computer technology, CHIEF is being developed for implementation in the early 1990s. CHIEF (Customs Handling of Import and Export Freight), is designed to introduce a comprehensive export system, capture real quota control data, link to the Department of Trade and Industry, and exchange data with the EEC in Brussels. A major benefit will be the production of better management information than at present. CHIEF will be capable of handling the Single Administrative Document (SAD) and the new tariff 'harmonised' system (HS) (see Glossary).

TRANSPORT AND SHIPPING

DISH/EDIS

The DISH (Data Interchange for Shipping) pilot project was commenced in October 1986, and involved linking the computers of 12 major exporters, freight forwarders and shipping lines. The firms involved included ICI, Guinness Exports, Rowntree Macintosh, OCL, Cunard Brocklebank, and ACT. The project used the TRADANET network service acting as a central 'clearing house'.

Five messages were designed, initially based on UN/TDI and UN/TDED standards. DISH has responded to EDIFACT developments, and in conjunction with DEDIST in Scandanavia, INTIS in Holland and SEAGHA in Belgium, has produced a harmonised set of messages which use the ISO 9735 EDIFACT syntax. The technical work of DISH and the maintenance and extension of the messages has now been passed over to the newly formed EDI Association. The new European shipping community, known as EDIS, is currently in the process of implementing these standards.

SHIPNET

The SHIPNET pilot project involved 40 companies, all IBM users, in the freight industry. Aimed at finance, government departments, customs and export guarantee agencies, SHIPNET documents con-

UK EDI APPLICATIONS

57

form to UN/TDI standards. The pilot project has now finished, and SHIPNET and DISH are now collaborating under the EDI Association to develop common EDIS message standards (see above). Due to conflicts with several other users of the same or a similar name, the continuing SHIPNET community is now known as the IBM EDI User Group.

CHEMICALS

CEFIC

CEFIC is a pilot EDI project organised by the European Council for Chemical Manufacturers' Federations. The project, supported by ICI, involves 15 member companies. Using GEISCO's EDI Express service, it is the first major EDI implementation to be based on the X.400 electronic mail standard, as well as EDIFACT. The pilot trials commenced mid-1988 and are scheduled to run for six months.

ELECTRONICS

EDIFICE

EDIFICE, Electronic Data Interchange for Companies with Interests in Computing and Electronics, was formed in 1986 as a European forum for the exchange of views, experiences and ideas on EDI. Its aim is to promote the use of EDI between computer manufacturers and their electronics suppliers for the exchange of business documents (orders, invoices, etc).

ECIF/AFDEC

This UK initiative for EDI stemmed from two leading trade associations, the Electronic Components Industry Federation (ECIF), and the Association of Franchised Distributors of Electronic Components (AFDEC). The three-year trial started in 1985, involves Philips Components, STC Electronic Services and Texas Instruments. The companies have used the TRADANET clearing-house service, with the association TRADACOMS standards.

HEALTH

The National Health Service places some four and a half million

58 WHAT IS EDI?

orders per year, and in return receives 12 million invoices and makes six million payments. As the NHS comes under increasing pressure to reduce costs, the potential for EDI developments within this sector is apparent, and both Istel and INS are active in the provision of EDI facilities for the Health Service.

Three Regional Health Authorities (Wessex, Trent, and North West Thames) use Istel's EDICT service to communicate with suppliers such as Kodak, Johnson and Johnson, and Grosvenor Surgical Supplies. The project has highlighted problems of standardisation in the coding system used to describe goods.

Wessex and Merseyside RHAs have been using INS's TRADANET service in pilot projects concentrating on pharmaceutical supplies. TRADANET also provides the PHARMANET specialist service to pharmaceutical manufacturers and wholesalers.

Another project run by INS links 20 general practitioners and two Family Practitioner Committees.

The NHS has set up a working party to recommend one EDI supplier for all 14 RHAs, and thereby solve the problems of authorities faced with subscribing to two independent networks.

BANKING

Banking has been at the forefront of automation since the late 1950s, as one might expect for a community enjoying extensive domestic and international branch networks, a large customer base, and a high volume of transactions. In particular, SWIFT and BACS are two well-established banking systems which deserve comment.

SWIFT

SWIFT, the Society for Worldwide Interbank Financial Telecommunication, has been running live for over 10 years. It carries up to a million messages a day, and is used by some 1500 banks worldwide. The project started with a remit of standardising banking transactions (international payments, foreign exchange deals, documentary credits, payments, etc) between different banks and has involved the maintenance and enforcement of interbank message standards, in addition to setting up the SWIFT worldwide telecommunications network. Whilst the membership remains exclusively to banks,

UK EDI APPLICATIONS

participation has now been broadened to include stockbrokers, securities exchanges and securities clearing and settlement institutions.

BACS

BACS, the Bankers Automated Clearing Service, is a Limited Company owned by the major UK clearing banks and a few other financial institutions. It was formed to provide an efficient, automated clearing service between participating financial institutions. Traditionally, input to BACS has been via magnetic tape and disk; however, there is now available telecommunications input via 'BACstel'. BACS is the largest automated clearing facility for electronic funds transfer (EFT) in the UK, with some 22,000 bank-sponsored users and a further 17,500 users bank-sponsored via BACstel.

INSURANCE

EDI activity in the insurance industry centres around three projects. LIMNET, the London Insurance Market network, will run on the IBM Managed Network Service, and plans to link over 1100 of Lloyd's underwriting agents, syndicates, brokers and insurance companies. RINET is a project created by the European re-insurance companies, and it has chosen IBM as its network service supplier. INS's BROKERNET service includes Commercial Union, Norwich Union, Zurich Insurance and Eagle Star. The network is used to exchange private motor car insurance details using industry specific data messages.

CONSTRUCTION

EDICON

EDICON, Electronic Data Interchange (Construction) Ltd, is an association of construction companies, including contractors, suppliers, manufacturers, architects, quantity surveyors and builders' merchants. It was formed in March 1987 with founder members including Trafalgar House, John Laing, Redland, Boulton & Paul, and the Graham Group, and expects 5000 members by the early 1990s. EDICON has identified the following areas for the application of EDI:

60   WHAT IS EDI?

– the manufacture, storage, sale and delivery of products;

– the selection and specification of products;

– the costing, ordering and payment for goods and services;

– the technical direction and control of the building process.

EDICON has formed development groups to investigate such business areas as invoicing, orders and enquiries, bills of quantity, product information, etc, as well as support groups covering industry coding, legal issues and standards.

TOURISM

UNICORN

UNICORN is the first of many expected EDI projects in the tourism and travel industry. It is an acronym for 'United Nations EDI for Co-operation in Reservation Networks'. The UNICORN project has concentrated on car ferry reservations and ticketing, and has involved three major passenger carriers in Europe: Sealink, North Sea Ferries, and P&O European Ferries (formerly Townsend Thoresen). The aim of UNICORN has not been to simplify paperwork (essentially the only piece of paper required is the passenger's ticket), but rather to apply EDI concepts in order to provide a common standard for processing interactive enquiry-type transactions. These transactions cover requests for and offers of ferry travel reservations, travel document production, pricing, and raising the resultant financial account. The project has given high priority to the setting of common standards, initially adopting and modifying the UN/TDI standard to cater for the requirements of real time conversations. Future implementations of UNICORN will adhere to the EDIFACT standard.

Although the system is presently confined to passenger reservations, its flexibility lends itself to possible future interconnections involving airlines, car hire companies, hotels, and HM Customs and Excise.

10 International Issues

Chapter 9 provided an overview of many of the projects significant in the development and current status of EDI within the UK. Since much of the UK business is carried out in the international arena, it is only to be expected that projects such as ODETTE, CEFIC, EDIFICE, EDIS and UNICORN could equally well be described under a chapter heading of 'International EDI Projects'. It is not the intention of this chapter to repeat or expand on any of these projects, but rather to discuss some of the related European initiatives that could have significant impact on longer term EDI developments.

1992

1992 heralds the arrival of a single European marketplace with the removal of barriers to European trade, including paperwork, varying taxes and differing national standards. The effects of these changes are likely to be far reaching, from the simplification of customs controls to the adoption of a common electrical 'Euro plug'. In a report published in March 1988 by the 'Office for Official Publications of the European Communities', a survey of 11,000 European businessmen showed that organisations perceive the greatest benefits of '1992' to be achieved from the removal of administrative barriers. It is therefore hardly surprising that many advocates of EDI are insistent that the single European market will not be feasible without the widespread use of electronic data interchange.

EUROPEAN INITIATIVES

SAD

SAD is the Single Administrative Document for all import and

61

62 WHAT IS EDI?

export transit and customs documentation in all EEC countries. The document was introduced on 1 January 1988 to replace virtually all of the customs freight documentation forms for export, import and transit movements within the EEC. The SAD was the most visible example of a set of changes called Customs 88, which included a new system of tariffs (see TARIC below) and the development of computerised procedures (see CD below). From an EDI point of view SAD is important since it provides the opportunity for the production of a standard customs freight documentation message. This gains even more significance when one considers that the document has also been adopted by EFTA and it is conceivable that it could become the basis for world-wide standard customs declaration.

CADDIA/CD

The European Community Commission CADDIA initiative, standing for 'Cooperation of Data and Documentation for Imports/Exports and Agriculture', has been running for over three years. The project is concerned with the use of telematics for EC systems concerned with imports, exports, and the management and financial control of the Agricultural Market organisation.

The formal vehicle for progressing CADDIA Customs Sector developments is the CD project (Coordinated Development of Computerised Administrative Procedures), which is divided into six main areas:

- intra-community trade;

- third country imports/exports;

- trade interfaces;

- member state – commission interfaces;

- communications; and

- data exchange standards.

The CD project team has been active in the development of the EDIFACT standard, and is seen as having a major role to play in

INTERNATIONAL ISSUES 63

facilitating the completion of the 1992 Internal European Market by the increased use of EDI.

TARIC

TARIC is the European Community Integrated Tariff (Tariff Intégre Communitaire). It allows nearly all goods subject to Community customs regimes (eg licensing, quotas) to be uniquely identified.

COST 306

The COST 306 project has concentrated on developing messages relevant to transport operators. Backed by the UN and EEC, the project is looking at ways of proving that EDI is workable in complex cross-border exchanges of information, and not just for the domestic sections of international trade. The project falls within the COST (Co-operation in Scientific and Technical Research) programme, designed to initiate and promote European collaboration in a number of high-technology areas. Participants in COST 306 include Austria, Belgium, Denmark, Finland, France, Italy, the Netherlands, Norway, Sweden, Switzerland, the United Kingdom, and West Germany.

LOG

LOG, Logistical Optimisation of Goods transport, is, like COST 306, a transport sector project, operating within West Germany at national level. The project has been concerned with differences in transport messages, hardware, and data communications. The LOG solution, which began operational testing at the end of 1986, is based on the use of front-end processors to build a unique interface to the existing hardware of each trading partner. LOG has started to look at the air cargo industry, and anticipates being finished by the first half of 1989.

DEDIST

The DEDIST (Data Element Distribution in Trade) project is a Scandinavian project, initiated by Finland and now including Denmark, Norway and Sweden. DEDIST is preparing guidelines for the representation of trade data in inter-Nordic and other international EDI communications.

INTIS

INTIS is a Rotterdam community project, founded in 1986 by the Rotterdam port authority and the Dutch PTT. The importance of Rotterdam as one of the world's major ports and a centre of commercial activity has given significant momentum to the project. The EDI services are seen as benefiting Rotterdam's customers and as strengthening its position as a gateway to Europe. INTIS uses internationally accepted standards for EDI messages, and collaborates with DISH, DEDIST and SEAGHA who share similar interests. In the international arena, the first gateway to a port information system will be between INTIS and the Felixstowe FCP80 system, using the IBM-INS network, and operational in the middle of 1988.

SEAGHA

SEAGHA is an acronym standing for Systems Electronic and Adapted Data Interchange in the Port of Antwerp. The company was formed in 1986 by a group of six professional associations in Antwerp, representing shipowners, shipping agents, freight forwarders, cargo handlers, stevedores and port operators. The SEAGHA EDI system includes support, a bridge interface to existing computer systems, the network and clearing house, and a manual detailing procedures, regulations and standards. Pilot operation is due to commence in the first half of 1988.

KOMPASS

KOMPASS was established in 1981 as the port of Bremen's integrated information and documentation system. The system handles 47,000 transactions a day for 96 companies. Specific port activities have their own communications systems. LOTSE supports transport handling, ship documentation and container control for companies engaged in maritime trade. DAVIS provides a container management system, linking forwarding agent, packing company and cargo handling company with the inland companies involved in the project.

DAKOSY

DAKOSY performs a similar function to that of KOMPASS, but for the port of Hamburg. The system was established in 1983, and

INTERNATIONAL ISSUES

65

currently supports about 64 forwarders, 22 liner agents and eight tally companies. As with KOMPASS, DAKOSY supports specific trade-oriented EDI sub-systems:

- SEEDOS for forwarders in seaports.

- TALDOS for tally/cargo control.

- CONDICOS for container control and scheduling.

- CONTRADIS for container transport and scheduling.

TEDIS

TEDIS, Trade Electronic Data Interchange Systems, is an action plan which has been drawn up by the Commission of the European Communities. The objectives of TEDIS are:

- To avoid a proliferation of closed community EDI systems and the consequent incompatibility.

- To promote the creation and establishment of trade EDI systems which meet user needs, particularly for small and medium-sized enterprises.

- To increase awareness amongst the European telematic equipment and services industry to meet user's requirements in this area.

- To support the common use of international and European standards.

TEDIS is keen to promote the transfer of information gained from specific sectoral projects, so as to achieve harmonisation on such common issues as standards, tariffs, confidentiality, security, etc. The project represents a commitment by the Commission to the development of the electronic transfer of trade data within the internal Community market.

11 Security and Legality

The preceding chapters have set forward the foundations of EDI, explaining the concepts, implementation issues and technical considerations. EDI standards, communications network services, and enabling hardware and software are all necessary building blocks to establishing the feasibility of EDI. However, just as necessary is the formation of a secure framework within which trading can operate. Whether trading is carried out by traditional paper-based methods or electronically, the requirements and features of this framework are likely to be the same.

There must be a secure method of transferring the information, which retains the ability to authenticate the source of the information, as well as giving evidence of the trading transaction. These prerequisites may be of a commercial nature, requiring new forms of contract, or of legal origin, requiring changes in the law. The latter requirement is certainly more difficult to resolve, since the law is paper-based and constructed on a foundation of well-known and well-defined problems. EDI, being a new technology, must of necessity formulate interim solutions until the law can 'catch up'.

The legal and security problems are apparent even in a simple one-to-one, single network trading link. They are compounded in the event of multi-partner, multi-network trading. It is not possible for such a small publication as this to address these problems in any depth. However, it may be instructive to examine some of the typical questions that arise in this area, particularly from the viewpoint of the receiver of EDI transmissions.

HOW CAN I BE SURE OF THE SENDER'S IDENTITY?

Standard paper documents containing signatures convey an authority

68 WHAT IS EDI?

which may at first sight be difficult to replicate with electronic transmission media. However, there are well-documented cases of the fraudulent use of paper documents. Well-managed computer networks offer a range of facilities to minimise the chance of unscrupulous users gaining access to EDI. These range from traders being able to register those companies with whom they wish to trade and the types of document supported (thereby also pre-empting the possibility of electronic junk mail!), to the use of usernames and sophisticated password checking mechanisms. In addition, the message standards themselves contain 'header' information, giving sender identifier and password, which may be coded or encrypted if desired.

CAN I BE ASSURED OF THE INTEGRITY OF THE MESSAGE?

The recipient must be certain that he has received all of the parts of the message, and in the right order. The EDI infrastructure contains a series of built-in checks and safety features. Modern communications protocols such as X.25 automatically check the sequence of the packets of data and also ensure that all the packets arrive at their proper destination. Network control and management software is designed to ensure that messages are stored and transmitted in the correct sequence, and is built to be resilient to line or other transmission failures. Lastly, as with the previous question, the EDI standards have been designed to allow for checks to be made at the start and end of a transmitted message.

IS THE TRANSMISSION MECHANISM SECURE AND CONFIDENTIAL?

Electronic mailboxes must be secure so that rival organisations, who may also subscribe to the same network, cannot inadvertently or intentionally gain access to confidential and potentially damaging information. Certainly the alternative transmission mechanism, ie the postal service, is not infallible in this respect. However, widely-reported cases of school children 'hacking' into supposedly secure computer databases give cause for concern to the organisation that entrusts its lifeblood to a third-party operator. These network operators must demonstrate that security has been a key feature in the design of the network, in addition to displaying an adequate level of physical security (access to staff, software and equipment).

SECURITY AND LEGALITY

69

Regular audits of the facility should be carried out by a reputable auditing firm.

Additional security may be established by encrypting the message. However, although this is technically relatively easy to do, albeit at extra cost to the trading partners, operational requirements and practice may dictate that the messages are sent 'as is'. Network operators and telecommunications administrators may prefer unencrypted messages, for ease of handling and charging, whilst customs authorities prefer such messages for ease of vetting. Organisations may have to settle for partial encryption of sender identifier and password, or make use of 'digital signature' techniques.

HOW CAN THE SENDER BE SURE THAT THE MESSAGE ARRIVED?

Traders will need to satisfy themselves that the level of reporting is sufficient for requirements. Some communications protocols, such as the X.400 message handling service, can guarantee receipt. The network operators should supply summary reports, listing all transmissions sent/received over a given period of time. Finally, the message standard should support the implementation of acknowledgements. An Interchange Agreement (see below) between trading partners will establish the precise and acceptable meaning of an acknowledgement.

DOES THE EDI MESSAGE CONSTITUTE A BINDING CONTRACT?

Perhaps the most important function of a trading document is as a legally binding contract to a trading transaction. In the absence of a solid legal foundation built on case histories, trading partners and the network supplier must be content with establishing Interchange Agreements. These agreements will act as a code of conduct, laying down rules and responsibilities. Such agreements have been successfully enforced by close user-group communities. Building on this success, the UNECE International Chamber of Commerce (ICC) has formulated a set of Uniform Rules of Conduct for Interchange of Trade Data by Teletransmission (UNCID). These UNCID guidelines may be adopted and incorporated into Interchange Agreements, thereby simplifying the process of setting up such agreements whilst

WHAT IS EDI?

at the same time laying the foundations for recognised, legally supported commercial practice. These UNCID rules may be obtained from:

ICC (UK)
Centrepoint
103 New Oxford Street
London WC1A 1QB

WHAT IS THE LEGAL STANDING OF AN EDI TRANSACTION?

Just as certain paper documents must be retained for 10 years (for example, to satisfy legal requirements), so too must the equivalent electronic data. Although it is without doubt simpler and more efficient to store this data electronically, the method of referencing the data for ease of retrieval will need to be addressed.

Currently, there is a legal requirement for certain transactions to be supported by a paper document. The negotiable Bill of Lading in use in the shipping industry is one such document, which conveys rights of ownership. Replacement of this document by a legally equivalent EDI version will necessitate changes to the law, a process which can only be progressed quickly by commercial pressure from those parties with vested interests.

The acid test which determines whether EDI is legally robust is the admissibility of computer produced evidence in a court of law, should a dispute arise over a commercial transaction. In this respect current legal rules of evidence, for example the 1986 UK Police and Criminal Evidence Act, appear to be less than ideal. The gap of understanding between lawyers and computer specialists must be bridged. Equally important is the need for good auditing facilities which can be used to demonstrate that the messages in question were actually sent, and that the computer was operating properly at the time of the transmission.

CONCLUSION

Without doubt the law has lagged behind the rapid rate of technological advance of the computer and telecommunications industry. This has posed problems for the EDI community in establishing a demonstrably secure legal framework in which to conduct business. These problems are compounded when EDI traffic

SECURITY AND LEGALITY

crosses national boundaries, using multiple networks and communications standards.

However, operating within national legal structures and using Interchange Agreements bolstered by codes of conduct, EDI messages can be demonstrated to be both secure and legally acceptable. Practitioners of EDI are therefore finding ways to overcome the legal obstacles; significant amongst these efforts is the work of the legal groups within ODETTE and the EDI Association. As pressure grows from the EDI trading community one can expect the legal framework to become more comprehensive and robust.

12 Conclusion and Future Developments

The EDI race is now well and truly underway! The UK and the USA are running neck and neck, with the rest of Europe trailing some 18 to 30 months behind. The exponential growth path shows no sign of flattening out. Current projections suggest that by the year 1995 400,000 companies world-wide will be trading using EDI, with Europe accounting for about 35% of this total, the USA 50%, and Japan and South East Asia providing most of the remainder.

It is perhaps surprising that this method of doing business has been so readily adopted by so many companies. Certainly the international EDIFACT standard lacks a foundation of user experience, and the long-term future of closed-community standards, under pressure to move towards EDIFACT, seems uncertain. The vendor market has been in many cases slow to respond, when one considers that there is still no interconnection between networks and that EDI packaged software is only just beginning to flourish. Perhaps the most serious hurdle preventing the EDI race from commencing should have been the uncertain legal status of this form of trading, stemming from the failure of the legal profession to keep pace with Information Technology developments.

However, these hurdles have not proved insurmountable; on the contrary, ways have been found round them, over them, and in some cases they have been knocked down completely. Established trading communities now exist using tried and tested software and exchanging information via industry agreed standards. Users have been prepared to subscribe to several networks with the consequent costs and complexities. Interchange Agreements set up between trading partners have provided an acceptable degree of legal security to EDI.

74   WHAT IS EDI?

In addition, inter-governmental and non-governmental international and national organisations have collaborated, through the ICC, to develop a legal foundation stone for EDI in the UNCID guidelines. Far from being intimidated by these hurdles, the EDI players have been prepared to join forces to place pressure on the supplier sector and the legal profession in order to effect the necessary changes. This pressure is also being placed on other members of the trading community, as many companies start to insist that their partners support EDI as a requirement for continued trading.

Put quite simply, companies are not prepared to sit back and wait for EDI to happen; the perceived benefits are too important. Simplification of administrative procedures and reduced paperwork alone can mean significant improvements in year-end profits. However, when one considers that EDI can strike right at the heart of business strategy (facilitating the establishment of just-in-time manufacturing techniques, and providing the speed of response so necessary to meet crucial customer service levels), the only catalyst necessary is often 'awareness'. In the trading sector recent estimates suggest cost savings of between 3.5% and 7% of transport cost and up to 10% of delivered consignment values, which may amount to $350bn annually, world-wide.

What does the immediate future hold for EDI? Already network vendors are starting to collaborate to link their proprietary services together – Istel and INS have announced such a link for 1988. The value of truly international standards is recognised and it is significant that the prestigious CEFIC project which commenced in 1988 within the European chemical industry has adopted EDIFACT. EDI, once the preserve of the motor and freight industries, is now moving into many new business sectors: the financial services sector (RINET and LIMNET are two recent examples), tourism and leisure, and health are all new areas of EDI growth, and we can expect the trend to continue.

Chapter 3 provided a definition of EDI and attempted to clarify this by pointing to differences and similarities between related technologies. In the future we can expect such facets of the office environment as electronic data interchange, electronic mail, CAD/CAM, FAX, and electronic funds transfer (EFT) to converge into a single integrated office. Already work is underway in the standards

CONCLUSION AND FUTURE DEVELOPMENTS

arena at using the X.400 Message Handling Guidelines as the basis for such an integration.

The landmarks are already on the horizon. 1992 should see the beginning of the single European market, with the removal of EEC trade barriers. EDI will be essential to reap the rewards offered by breaking down these restrictions. The introduction of the Integrated Services Digital Network (ISDN) will represent an immense leap forward in the capacity for information transmission, providing faster transfer speeds, increased geographic coverage, and a much greater degree of data integrity. For EDI, which has been built around information transmission, the significance is self-evident.

EDI has already firmly established its place within many business strategies. The future seems certain to leave no aspect of industry and commerce untouched: the question is not 'if?' but 'when?'.

Appendix 1

Glossary

ANA

Article Number Association. The ANA is a trade association funded primarily by the retail trade and responsible for developing the familiar bar coding system seen on many UK products. It developed the TRADACOMS EDI standard and sponsors INS's TRADANET value added and data service.

ANSI/ANSI X12

American National Standards Institute. ANSI X12 is the dominant US standard for message structures, syntax and data elements.

ASTI

Association des Services Transports Informatiques. A cheap and cheerful service for freight forwarders, using the international Dialcom network (eg Telecom Gold in UK).

BACS

BACS, the Bankers Automated Clearing Service, is a consortium of the major UK clearing banks, which was formed to provide an efficient, automated clearing service between participating financial institutions.

BROKERNET

A VAD service, supplied by INS, for high street insurance brokers.

77

CADDIA/CD	CADDIA, standing for 'Cooperation of Data and Documentation for Imports/Exports and Agriculture', is a European Community Commission initiative. The project is concerned with the use of telematics for Community systems relating to imports, exports, and the management and financial control of the Agricultural Market organisation. The formal vehicle for progressing CADDIA Customs Sector developments is the CD project (Coordinated Development of Computerised Administrative Procedures).
CHIEF	The Customs Handling of Import and Export Freight, is a UK Customs EDI system which is being developed for implementation in the early 1990s, as a replacement for DEPS. The system is designed to introduce a comprehensive export system, capture real quota control data, link to the Department of Trade and Industry, and exchange data with the EEC in Brussels.
COMPAT	A computer aided trade conference, held annually and supported by the Commission of the European Communities.
COST 306	An organisation, backed by the UN and the EEC, which has concentrated on developing transport messages relevant to complicated cross-border interchanges in international trade.
CUSTOMS 88	A project to introduce a single, internationally agreed commodity classification, known as the Harmonised System (HS), TARIC, SAD and CHIEF.

GLOSSARY

Data Elements Individual fields (items) in an EDI message.

DEDIST A Scandinavian EDI group, promoting EDI across the Nordic countries.

DEPS A UK Customs system for handling electronic entries, and standing for Departmental Entry Processing. DEPS is to be replaced by CHIEF in the early 1990s.

DISH Data Interchange for Shipping. The DISH project comprises a group of UK exporters, freight forwarders and shippers, who have completed a successful pilot EDI trial, using the TRADANET service. The technical work of DISH has since been passed over to the EDI Association, whilst the DISH community is now working with other similar European groupings under the new name of EDIS.

DTI Direct Trader Input to HM Customs. DTI is the method by which carriers and agents input information on-line to the Customs' DEPS computer system. All validation queries are sorted out before the Customs personnel become involved, providing fast clearance of Customs entries. DTI is in use in ports such as Southampton and Felixstowe.

ECE Economic Commission for Europe. A United Nations organisation based in Geneva and responsible for the development of the EDIFACT EDI standard.

ECIF Electronic Components Industry Federation, which has completed an EDI trial.

EDI	Electronic Data Interchange. The transfer of structured data, by agreed message standards, from one computer system to another, by electronic means.
EDI Association	A UK association of over 150 members formed in September 1987 to promote and develop EDI in international trade and related industries. Specialist sections cover air, land and sea transport, banking, insurance, customs and government. The association is a member of the European EDIS message development group.
EDICON	An EDI community for the construction industry, whose members include contractors, suppliers, manufacturers, architects, quantity surveyors and builders' merchants. It was formed in March 1987 and expects 5000 members by the early 1990s.
EDICT	An EDI value added and data service provided by Istel.
EDI*EXPRESS	GEISCO'S international EDI network service. In the UK this is being linked with TRADANET as part of the INS joint venture with ICL.
EDIFACT	Electronic Data Interchange for Administration, Commerce and Transport. EDIFACT is an international standard agreed between UN-ECE and ANSI, which provides syntax rules, segment construction and transmission rules to enable EDI message structures to be formulated.
EDIFICE	The Electronic Data Interchange for Companies with Interests in Computing and

GLOSSARY

81

Electronics, was formed in 1986 as a European forum for the exchange of views, experiences and ideas on EDI. Its aim is to promote the use of EDI between computer manufacturers and their electronics suppliers for the exchange of business documents (orders, invoices, etc).

EDIS

EDIS is a transport messages working group, involving shipping communities from the UK EDI Association, the Scandinavian DEDIST group, INTIS from the port of Rotterdam, and SEAGHA from the port of Antwerp.

Equator

A software package developed for microcomputers by INS, and designed to make it easy for shippers to create, send and receive EDI messages.

Euromatica

A Brussels-based consultancy, which organises the COMPAT conferences, and publishes a Europe-wide EDI newsletter called *Tradeflash*.

FASTRAK

A value-added and data service provided by the Midland Bank.

GEISCO

General Electric Information Services Company. GEISCO is part of the General Electric Company and provides a value added and data service running on its world-wide network.

HS

The Harmonised System. A 14-digit system of commodity codes, enabling goods and their origin to be identified anywhere in the world.

IDEA	International Data Exchange Association, based in Brussels.
INS	International Network Services. A joint venture company formed between ICL in the UK and GEISCO in January 1987. INS has 'brand-named' its (value added and data) EDI service to the business sector, providing TRADANET, MOTORNET, BROKERNET, DRUGNET, and PHARMANET.
Interbridge	A software product developed by SITPRO to handle the formatting (construction) and deformatting (translation) of EDI messages. Interbridge supports the UN/TDI and EDIFACT message standards.
INTIS	The EDI community in the Port of Rotterdam.
ISTEL	A leading UK system house which provides the EDICT value added and data EDI service.
JEDI	The Joint EDI Committee of the UN and ANSI. This task force has brought together the EDI standards of America and Europe into the EDIFACT standard.
LACES	London Cargo EDP Scheme. This freight clearance system for Heathrow Airport ran from 1971 to 1981.
LIMNET	The London Insurance Market Network. An EDI user grouping, headed by Lloyd's, and using IBM's Information Exchange for providing the communications service.

GLOSSARY

Mailbox
An area within a VAD service where data is deposited by the network for subsequent collection by the owner of the mailbox.

Message
In simple terms a message corresponds to a trade document. It is a specific EDI structural unit, consisting of a group of segments brought together for a specific purpose and sent electronically.

MNS
Managed Network Services. IBM's value added and data service offering, which provides an EDI service in Information Exchange.

ODETTE
The Organisation for Data Exchange by Teletransmission in Europe is the EDI group for the motor industry, linking vehicle and component manufacturers in eight European countries. ODETTE has developed its own message standards based on UN/TDI.

OSI
Open Systems Interconnection. A model for designing computer systems which allows the computer systems made by different manufacturers to communicate with each other.

Period Entry
This procedure allows for the monthly transmission of statistical and accounting information to customs, via magnetic tape or direct transfer. (Contrast with DTI.) Information is produced by the company's own accounting and stock control systems. Period Entry is mainly carried out on the importing side at present.

PHARMANET
INS's value added and data service for the pharmaceutical industry.

84 WHAT IS EDI?

Postbox

The area within a VADS network into which a user sends all his transmissions.

RINET

This Re-insurance and Insurance network is a recently formed European EDI community.

SAD

The Single Administrative Document. Effective from 1 January 1988, this document replaces about 100 customs declaration forms in use within the EEC. It is to be adopted by EFTA countries also, and may eventually become the standard for a world customs declaration for freight.

Segment

A logical grouping of data elements.

SHIPNET

The group of IBM users in the shipping industry, which has completed a pilot EDI project. This group is now known as the IBM EDI User Group.

SITPRO

Simpler Trade Procedures Board. A British organisation set up by the Department of Trade and Industry to rationalise international trade procedures. SITPRO has been instrumental in the development of trade data standards, and in encouraging British Industry to adopt EDI.

SPEX

A microcomputer software package, developed by SITPRO, that stores and processes international trade data and produces export invoices and other documents.

Store & Forward

The service offered by a VADS supplier which allows the sender and receiver of an EDI transmission to be 'decoupled'.

GLOSSARY

85

The data is forwarded to a network post-box at the convenience of the sender, stored on the postbox/recipient mailbox prior to being forwarded, at the convenience of the recipient, to his computer system.

Sub Elements

The smallest item of data included within a segment.

SWIFT

The Society for Worldwide Interbank Financial Telecommunication. A banking network that has been running live for over 10 years, carries up to a million messages a day, and is used by some 1500 banks worldwide. SWIFT members also include stockbrokers, securities exchanges and clearing and settlement institutions.

Syntax

Standards for the construction and transmission of EDI messages.

TARIC

The European Community Integrated Tariff (Tariff Intégre Communitaire). It allows nearly all goods subject to Community customs regimes (eg licensing, quotas) to be uniquely identified.

TDED

The Trade Data Elements Directory issued by the UN (ECE). It provides the vocabulary for constructing EDI messages.

TDI

Trade Data Interchange. Term that used to be used for EDI.

TEDIS

Trade Electronic Data Interchange Systems. An action plan drawn up by the EEC to popularise EDI, particularly to promote international standards, and to avoid a

86 WHAT IS EDI?

proliferation of incompatible closed-community systems.

TRADACOMS

A UK EDI standard, developed in 1982 by the ANA, based on UN/TDI. The initial tender to provide a third-party VADS bureau supporting TRADACOMS was won in 1983 by ICL and its TRADANET service. TRADACOMS is now the most widely used EDI standard in the UK, and is supported by all the major VADS suppliers.

TRADANET

The INS value added and data service for the manufacturing, wholesale, distribution and retail sectors within the UK. TRADANET International is the parallel service for the international trade community.

Transmission

All the data to be transmitted between one user and another, or between one user and the VADS.

UNCID

A set of legal rules proposed by the International Chamber of Commerce to cover paperless transactions.

UNECE

United Nations Economic Commission for Europe. Based in Geneva, the UN-ECE has been instrumental in building on the standards work done by SITPRO and extending this to the European arena. They have produced a Trade Data Element Directory (UN/TDED) together with syntax guidelines for Trade Data Interchange (UN/TDI).

UN/TDED

The Trade Data Elements Directory issued by the UN-ECE, and providing standard

GLOSSARY 87

data elements for the construction of segments and EDI messages.

UN/TDI These guidelines for Trade Data Interchange, developed by the UN-ECE, define standard syntax for the construction of trade data messages. They have been used as the basis for data exchange by many industries and applications, mainly in Europe. Trade associations have adapted the TDI standards to produce versions for specific business sectors, eg TRADACOMS for the retail trade, ODETTE for the motor industry.

UNSM United Nations Standard Message. A message for a specific purpose, and accepted by the UN for that purpose.

VADS Value Added and Data Services. A third-party bureau service which provides not only the physical network for data transmission, but also ancillary facilities such as mailboxes, security, auditability and promotion. The VAD services are generally regarded as including EDI, electronic mail and on-line databases.

Vanguard A joint UK government and industry initiative to promote the use of VADS.

VANS Value Added Network Services. The name formerly used for VADS.

Appendix 2

EDI Groupings and Contacts

This appendix provides a brief description and contact address for some of the organisations involved in promoting and disseminating information on EDI.

EDI ASSOCIATION

The EDI Association (EDIA) is a UK grouping formed in September 1987 to promote the development of EDI in international trade and related industries. The main objectives are:

- to provide a UK EDI forum for international trade and transport activities;

- to promote and encourage the use of EDI by all participants in international trade;

- to promote the use of standards;

- to support the implementation of various EDI applications;

- to help foster the progress of EDI on a worldwide basis.

The EDIA is divided into specialist sections covering short sea, air transport, deep sea, insurance, banking and financial services, and customs and other government related issues. In addition specialist bodies cover message coordination (working with INTIS, DEDIST, COST 306 and SEAGHA), legal issues and communications.

90   WHAT IS EDI?

Contact:

1st Floor
Almack House
26 King St
London SW1Y 6QW

Tel: 01 930 0532

INTERNATIONAL DATA EXCHANGE ASSOCIATION

IDEA was launched in May 1987 as a non-profit association of organisations whose aim is the world-wide promotion, pursuance and application of Electronic Data Interchange. The Association brings together the interests of some of the most important EDI users in a group with substantial lobbying power and influence on policy-making. IDEA has set itself defined targets and functions including:

- the provision of facilities for meeting, consulting, and defining and promoting common views and opinions;

- to assist in the process of international standards recognition;

- to cooperate with other relevant organisations in promoting EDI through the support of appropriate standards and techniques;

- to mobilise international support for EDI policies;

- to accelerate and support activities for defining and implementing standard messages in specific industries;

- to focus and consolidate commercial and industrial interest at national and international level in the usage of EDI;

- to expand the association's activities world-wide;

- to support activities for defining, implementing and agreeing standard messages by establishing working parties in specific industries.

EDI GROUPINGS AND CONTACTS

IDEA members are drawn from across the EDI spectrum, including end-users, network vendors, suppliers, and other interested parties.

Contact:

68 Avenue d'Auderghem
Bte 34
1040 Brussels
Belgium

Tel: (32.2) 7369715

NATIONAL COMPUTING CENTRE

The NCC is a membership-based organisation whose aim is to promote the effective use of Information Technology. Membership is drawn from all parts of the IT community, with the NCC offering a collective view of the community, to the community. A prime focus of the Centre's work is on technology transfer, researching a topic area and providing the resultant outputs and services in the form of advice, seminars, training, reports, publications and consultancy. In all of these activities the NCC strives to operate in an impartial manner, independent of a particular supplier's products and services.

The NCC is particularly concerned with IT developments that affect user strategy and on achieving educated awareness by top management of the benefits of IT. In this respect EDI is seen as an important area of work, and the NCC has been active in researching the subject, hosting seminars, and producing publications and reports, as well as playing a full role on various standards development bodies.

Contact:

The National Computing Centre Ltd
Oxford Road
Manchester
M1 7ED

Tel: 061 228 6333 ext 2482

SITPRO

SITPRO, the Simpler Trade Procedures Board, is an independent body set up by HM Government in 1970 and sponsored by the Department of Trade and Industry. The Board draws its members from industry, commerce and government, and employs a total of 26 full-time staff. It works to reduce the cost of international trade by:

– developing relevant national and international standards;

– negotiating simpler procedures and information requirements;

– applying high technology solutions.

SITPRO has contributed to the development of many standards and simplified procedures used around the world. In 1978 it published the first set of EDI rules in Europe for the interchange of trade data, which went on to form the basis of the widely-adopted UN/ECE syntax guidelines for Trade Data Interchange (UN/TDI). SITPRO commissioned a software house to develop application software for these syntax rules, and now markets the resulting Interbridge package. Interbridge, together with the SPEX package for exporters, are now in wide use amongst the EDI community.

Contact:

Almack House
26 King Street
London SW1Y 6QW

Tel: 01 930 0532

VANGUARD

VANGUARD was launched in October 1986 as a joint government and industry initiative to improve the profitability and competitive position of British industry by promoting the increased use of Valued Added and Data Services (VADS). The initiative, led by the Department of Trade and Industry, has received the backing of five leading UK information technology companies: British Telecom, IBM, INS, Istel and Midland Bank.

EDI GROUPINGS AND CONTACTS

The first phase of VANGUARD, completed in May 1987, concentrated on promoting general awareness of VADS. In its second phase VANGUARD has worked with trade associations and other industry groups to identify communities of common interest and to help them towards implementation of VADS.

VANGUARD has commissioned several VADS-related studies, and has published the findings in the form of HMSO reports.

Contact:

Room 937 Kingsgate House
66–74 Victoria Street
London SW1E 6SW

Tel: 01 215 2627

Index

1992 (Single European Market)	61, 63, 75
ACP80	54, 55
ACP90	55
AFDEC	57
airports	55
ANA (Article Number Association)	44
ANS X12	43
ASTI	00
automobile industry	53
Background to EDI	3
BACS	59
BACstel	59
Banking	58
Benefits	
acknowledged receipt	15
awareness of	5
bargaining power	14
cashflow	15
competitive edge	25
cost savings	4
enhanced image	16
general	11
operational	12
opportunity	12
reduced costs	14
security	15

WHAT IS EDI?

speed	12, 13
stock holding	14
strategic	12
Bill of Lading	19, 70
British Telecom	35
BROKERNET	34, 59
CADDIA/CD	62
CEFIC	44, 57
chemical industry	57
CHIEF	56
clearing house	30
communications	27
implementation consideration	49
private networks	28
Public Switched Network	28
COMPAT	78
contruction industry	59
consultancy	50
consumer goods industry	54
COST 306	63
customs (see HM Customs and Excise)	
CUSTOMS	78
DAKOSY	64
Data element	41
composite	41
definition	79
directory	41
sub element	85
DEC	34
DEDIST	63
DEPS	55
DISH	46, 56
DRUGNET	34
DTI (Direct Trader Input)	54, 55
ECE (Economic Commission for Europe - see UN/ECE)	

INDEX

ECIF	57
EDI	
definition	7
example	23
future	73
planning for	48
EDI Association	56
EDI*EXPRESS	81
EDICON	59
EDICT	34, 53, 54, 58
EDIFACT	43, 46
EDIFICE	57
EDILINK	36
EDINET	35
EDIS	46, 56
electronic data interchange definition	80
electronic funds transfer	59
electronic mail	9
electronics industry	57
EQUATOR	36
EUROMATICA	81
Fastrak	35
FCP80	54
FORDNET	54
functional group	41
GEISCO	33
Health service	57
HM Customs and Excise	55
CHIEF	56
DEPS	55
Direct Trader Input	55
SAD	56
HS (Harmonized System)	81
IBM	33
ICL	33
IDEA (International Data Exchange	82, 90

Association)	
definition of EDI	7
Impact of EDI	16
cooperation	16
education	17
implementation considerations	48
methods of working	17
role of standards	16
third parties	17
implementation of EDI	47
Information Exchange (IBM)	33
Infotrac (Istel)	34
INS	33, 82
Insurance industry	59
interactive EDI	9, 60
Interbridge	36
Interchange agreement	9, 69
International Chamber of Commerce	19, 69
INTIS	64
invoices	24
ISDN (Integrated Services Digital Network)	75
ISO 7372 (Trade Data Element Directory)	43
ISO 9735 (see EDIFACT)	
Istel	34
JEDI (Joint EDI task force - see UN/JEDI)	
just-in-time	3, 14, 25
KOMPASS	64
LACES	3, 35, 55
legal aspects	19, 67
LIMNET	33, 59
LOG	63
magnetic tape	9, 27
mailbox	10, 30
Managed Network Service (IBM)	33

INDEX

99

MERCATOR	54
message	41, 83
design guidelines	41
MOTORNET	34, 53
NCC (National Computing Centre)	91
negotiable documents	19
ODETTE	43, 44, 53
OSI	83
PACE	55
paperless trading	7
period entry	83
PHARMANET	34
ports	54
postbox	30
production schedules	24
PSTN	28
RINET	59
Risk of adopting EDI	18
SAD (Single Administrative Document)	56, 61
SEAGHA	64
security	67
segment	41
self-billing	17
SHIPNET	56
shipping industry	56
SITPRO	42
SMMT	53
software	35
enabling service	32
implementation issues	49
packages	19
SPEX	36
standards	8, 39
considerations	49
development	42, 45

100 WHAT IS EDI?

example	42
future	45
importance	40
relation to trade data	39
SWIFT	58
syntax	
definition	85
rules	41
Systems Designers Ltd	36
TARIC	63
TDED (Trade Data Element Directory - see UN/TDED)	
TDI (Trade Data Interchange syntax -see UN/TDI)	
TEDIS	65
Telecommunications	
deregulation	4
revolution	4
tourism	60
TRADACOMS	43, 44, 54
TRADANET	33, 44, 54, 58
training	50
Travinet	34
UKAS	55
UN/ECE	43
UN/JEDI	43
UN/TDED	43
UN/TDI	43
UNCID rules of conduct	69
UNICORN	60
UNSM	87
USA	3, 43
VADS	
benefits	32
costs	35
definition	29
IBM	33

INDEX

INS	33
interconnection/interworking	18
Istel	34
services	30
suppliers	33
Vanguard	37
VANS	87
Waybill	19
X.400	75